解读台式新空间美学
INTERPRETING NEW SPATIAL AESTHETICS OF TAIWANESE STYLE

台风台韵

Taiwanese genre and charm

Shenzhen Design Vision Cultural Dissemination Co., Ltd
深圳视界文化传播有限公司 编

PREFACE / 序

Residence is the container of life; it reflects the owner's life attitude, aesthetic taste and cultural identity. It is no longer the ideological design in a spatial form, but also shows people's interpretation and practice of "home".

Spatial design in the 1980s Taiwan, China is concerned with the development of architecture and interior design, and the 1990s is influenced by the educational design in Taiwan. In the early period, designers paid much attention to environment, interior and architectural performance. But now, they are focusing on fashion, art, technology, environmental protection and health, showing the diversity of concepts and space. Over the past ten years, residential design in Taiwan and mainland showed the rich contemporary "Living Space", reflecting the conversion of their living environment and connotation. Designers in Taiwan are expert in building humanistic spaces, designers in Hong Kong are good at creating fashionable atmosphere and styles, while designers in mainland are skillful in showing various vision effects and exploring different styles.

In an era when the information of global design flows rapidly, the interior designs in China are still affected by Western aesthetics, with little innovations. Some works are trying to integrate Oriental elements, but they are merely with superficial performances and failed to reflect the deep value of traditional aesthetics. With the development of economy in China, many residential designs tend to be luxurious. And these projects are always concerned with the decoration of splendid materials, while less attention is paid to the consideration of human and environment. This is derived from the narrow vision of the homeowners' desire for presenting their wealth.

Chiang Xun, a famous Taiwanese artist once said, "All the designs will finally return back to life, which should be taken into consideration by all the Taiwanese designers." This gives a reflection for designers, that is, while presenting the styles and materials, designers should also explore the social life value and create spaces that conform to Chinese characteristics.

TWA design has been exploring residential designs in the past few years and always shoulders the responsibility of offering professional residences, reflecting the life attitudes and aesthetic tastes of the homeowners. Here, I take this opportunity and make an expectation that the future designers in Taiwan and mainland should pay much attention to the traditional cultural deposits, contemporary life aesthetics and environmental designs, so that they will present their unique aesthetic style conformed with their life patterns,

CARRYING FORWARD THE VALUE OF DESIGN AND USHERING IN THE LIFE ATTITUDE OF CONTEMPORARY CHINESE

发扬设计价值，引领当代华人生活态度

which differs from European, American and Japanese designs. Even the ordinary people will realize the value of design and the improvement of life, with such a design concept that designers will reflect the influence of society on life pattern.

住宅是生活的容器，反映着居者的生活态度、美学品位与文化特征，"住宅"不再只是空间形式的表象设计，也呈现出人们对"家"的定义与生活内涵的具体实践。

80年代的中国台湾空间设计者多是以建筑与室内设计养成背景为主，90年代则广受台湾设计教育兴起的影响，设计从早期关注环境、室内、建筑操作之课题，近年则普遍地与时尚、艺术、科技、环保、健康等领域整合，呈现出多元的思维操作与多样的空间风貌。综观近十载两岸各地的住宅设计发展，作品中展现了丰富的当代"生活空间"面貌，可看出华人生活环境与内涵的转变。台湾设计师精于塑造人文气质的空间场域，香港擅长开拓时尚氛围及风格营造，大陆则展现丰富多样的视觉效果与各式空间风格的探索。

在全球设计信息流通迅速的当下，当代华人住宅设计仍深受现代西方美学的影响，许多作品常有高度相似性，较缺乏独特创新的语汇。有些作品似乎尝试融入东方的元素，但却又流于符号表象的操作，未能转化展现华人传统美学的深层价值。随着两岸人民经济发展与富裕，许多的住宅设计不免走向奢华，但这些强调奢华的作品，常流于富丽建材之堆砌装饰，缺乏对人文与环境的关注，这可能源于业主对于财富的展现欲望或帝王贵族生活的狭隘视野。

台湾知名艺术家蒋勋曾说过："设计最终仍是回归、落实于生活，这一点，或许是在台湾这块土地操作设计的设计者，必须深思熟虑之处。"这点给了设计工作者一个反省观点，那就是设计者在急于展现自我空间风格创作与材质堆砌之余，也应探求当代华人社会生活价值，并形塑符合华人特色的空间原型。

传十设计近年深耕住宅设计领域，一直以建筑专业整合、重视展现屋主的生活态度与美学品位为职志。借着本文机会，期许未来两岸华人设计工作者，能更加关注于传统文化底蕴、当代生活美学与永续环境设计等课题，展现出有别于欧美日的设计潮流，开拓属于当代华人生活的美学风格！甚而透过设计，让一般社会民众体认识设计改造与提升生活的价值，并透过设计理念引领社会对生活型态的省思。

传十设计 / 许天贵建筑师、李文心设计师
TWA DESIGN / ARCHITECT: TIANGUI XU
INTERIOR DESIGNER: WENXIN LEE

CONTENTS / 目录

NATURAL STYLE | 自然风

- 008　A REFLECTION OF GOOD LIFE
美好生活，就是这样

- 034　EXTENSION OF SPACE
空间层理的衍续

- 052　THE AXIS OF LIFE
生活轴线

- 060　BEAUTY! CUSTOM-MADE HUMANITY SKYLINE
美哉！订制人文天际线

- 070　THE JOURNEY OF TIME
时间旅程

- 080　DIALOGUE AND FRAME
对话·框景

- 090　SENSITIVE RESIDENCE ON THE RIVER BANK WHERE WATER JOINS THE SKY
水天一色的河岸疗愈宅

- 100　CUSTOM-MADE GRAND RESIDENCE
量身订制大气私宅

- 108　A PLEASANT RESIDENCE WITH APPROPRIATE ARRANGEMENT
统整脉络·敞心居所

- 116　A HOME WITH AXIS AND SCENERY
轴向·引景 家的聚场

- 124　LEISURE BLACK FOREST RESIDENCE
自然黑森林慵懒宅

LUXURIOUS STYLE | 奢华风

- 134　LUXURY BARN LIVING
奢华谷仓生活

- 150　THE BEAUTIFUL RURAL RESIDENCE
乡林景居

160	**IDEALISM AND ARBITRARINESS** 唯心·随意	212	**AN ARTISTIC STATE** 艺境
170	**GO HOME OR TRAVEL** 回家·旅行	220	**MOVEMENT AND CONCERTO** 乐章·协奏曲
182	**TRANQUILITY** 静谧	228	**A RESORT-STYLE RESIDENCE** 度假式住宅
190	**ENJOY LUXURY** 尊享奢华	236	**THE ELEGANT RESIDENCE IN A HOTEL STYLE WITH EMOTIONAL LIGHTS** 日光叙意 饭店式的品位居宅
204	**A GORGEOUS NEW YORK STYLE RESIDENCE** 华丽的纽约		

INDUSTRIAL STYLE | 工业风

248	**GATHERING** 聚	292	**TRUE HOME** 身心居所
266	**SOJOURN** 旅居	300	**FLEETING TIME** 光影流年
274	**THE WALL** 一道墙	310	**THE WHITE RESIDENCE** 白色大宅
282	**THE BEAUTY OF MINIMALISM** 极简之美		

室内绿化 INTERIOR VIRESCENCE

自然采光 NATURAL LIGHTING

装饰材料 DECORATIVE MATERIALS

设计理念 DESIGN CONCEPT

色彩 COLORS

引景入室 BRINGING SCENERY INTO HOUSE

空间规划 SPACE PLANNING

NATURAL STYLE

PRESENTING RUSTIC AND HUMANISTIC DIMENSIONAL
DEPOSITS WITH MATERIALS CLOSE TO NATURE

自然风

以贴近自然的素材,展现质朴、人文的空间底蕴。

NATURAL STYLE — PRESENTING RUSTIC AND HUMANISTIC DIMENSIONAL DEPOSITS WITH MATERIALS CLOSE TO NATURE

A REFLECTION OF GOOD LIFE

美好生活，就是这样

Location ǀ Taichung, Taiwan 项目地点 ǀ 台湾台中	**Area** ǀ 435m² 项目面积 ǀ 435m²
Designer ǀ Qingping Zhang 设 计 师 ǀ 张清平	**Design company** ǀ TIENFUN Interior Planning Ltd 设计公司 ǀ 天坊室内计划
Main materials ǀ marble, wood, leather, etc. 主要材料 ǀ 大理石、木质、皮革等	

DESIGN CONCEPT ǀ 设计理念

We believe this is what we call "good life".

In the open space, there is a flow of light which emphasizes a beautiful aesthetic living. With seemingly no deliberate design, the designer uses materials with memorable marks, and some details are coupled with historical imprint. Simple but bright room is themed with the understanding of the pace of life, and this unique perspective satisfies various potential life demand. A compassable, expansive, interactive and communicative space can motivate new experiences.

我们相信美好生活，就是这样的！

开阔的空间里有流动的光影，强调一种美而好的生活美学概念，用看似不刻意的设计，娴熟地使用了带着记忆符号的材质，一些细节有着沉淀的历史印记。简单而光线充沛的自然感的空间，通过理解生活的节奏为主题，以独特的视点，满足生活者各种潜在的需求。可围绕、可展开、可互动、可交流让空间与人不断激发新的体验。

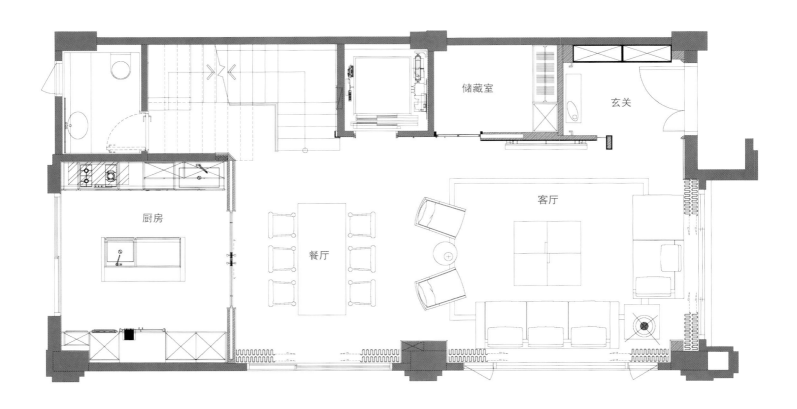

DECORATIVE MATERIALS | 装饰材料

Every small part and decoration as well as the selection of materials is creating craftsmanship that can not be easily achieved to make a comfortable living space. Paying much attention to the details of life, the design team uses montage to create a good life with aesthetics; each part is full of gratitude and treasure for good life.

　　每一个细小的局部和装饰，以及每一样选材，都在为舒适生活深思熟虑，创造一种极不容易达到的精工细作。看似自然、当然的处理其实凝结着对于生活丝丝入扣的用心以及设计者对于使用者考虑周全的智慧。设计团队以关注细节的生活提案，通过蒙太奇剪辑手法，开启一种感觉良好的美学生活，对于美好生活的感恩与珍惜浸润在每一个角落。

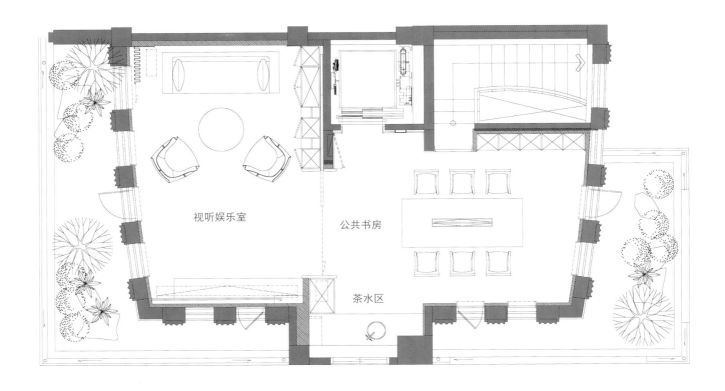

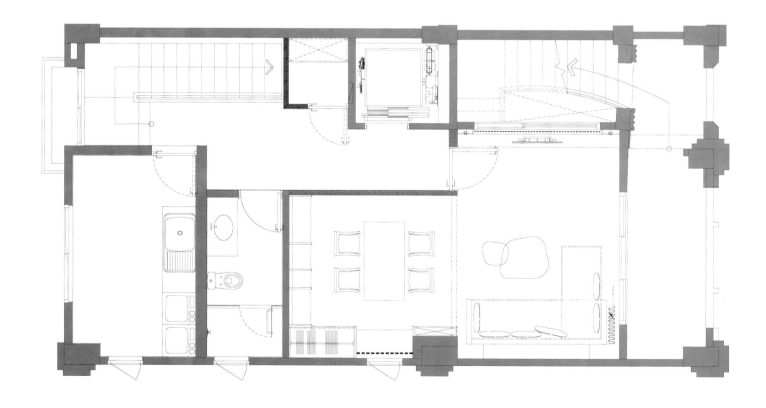

COLORS | 色彩

Good life derives from the appreciation and possession of all the wonderful things. The grayish marble wall with restrained and humanistic feeling matches perfectly with the L-shape gray leather sofa. The match of white marble table and the light brown leather chair is fashionable. The glittering crystal chandelier is luxurious and distinctively shaped. A few green plants and flowers add some romanticism.

美好生活的开始源于对美好事物的欣赏与拥有。浅灰色全质感大理石墙壁，低调内敛又富有人文气质，与灰色皮革L型沙发相辅相成。而白色大理石餐桌与浅棕色皮革椅的搭配，时尚亮眼。闪闪发亮的水晶吊灯，造型前卫奢华。几株绿植与鲜花的摆设，增添了更多浪漫生活的气息。

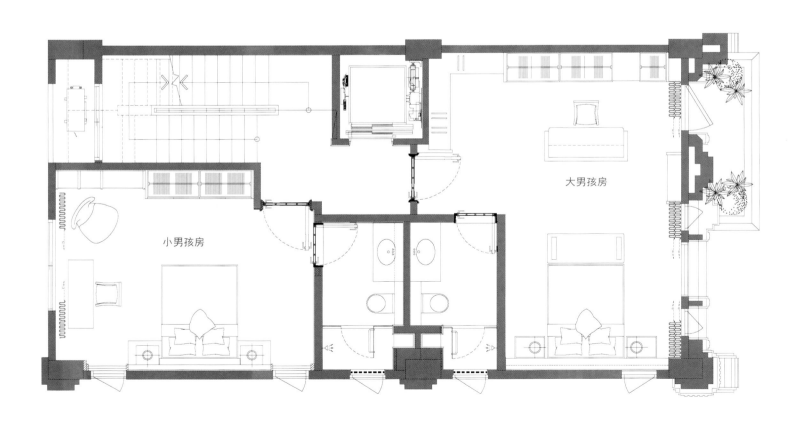

NATURAL STYLE
PRESENTING RUSTIC AND HUMANISTIC DIMENSIONAL DEPOSITS WITH MATERIALS CLOSE TO NATURE

EXTENSION OF SPACE
空间层理的衍续

Project name ｜ Milan Town 项目名称｜米兰小镇	Location ｜ New Taipei City, Taiwan 项目地点｜台湾新北	Area ｜ 330m² 项目面积｜330m²
Designer ｜ TT 设 计 师｜唐忠汉	Design company ｜ DESIGN ARTMENT 设计公司｜近境制作	Photographer ｜ Figure x Lee Kuo-Min Studio 摄 影 师｜图起乘李国民摄影事务所

Main materials ｜ wood veneer, iron part, stone husk, stone, etc.
主要材料｜木皮、铁件、石皮、石材等

DESIGN CONCEPT ｜ 设计理念

"The whole world is a stage,
And all the men and women merely players.
They have their exits and their entrances."
-From *As You Like It* by William Shakespeare
Just as the competent role of space the designers have made.

Cast/ Stairs
Link fields of layers through the linear connections,
Between shadows and buildings there flow with line.

Cast/ Living Room & TV Wall
Mass division with material extensions, simplicity of Humanity.
Tranquil light with harmony as the time flows.

Cast/ Tearoom
Combination of old piles,
Strengthen space mass of structure.
Through the boards the lights of orders,
Where balance of Yin and Yang is found.
Functional extension with silence lying on the edge of scene,
Broaden views with sitting heights.
Flowing wit simplicity, low-key and Zen spirit.

Cast/ Light Wall, Garage & Air Bridge
Through the light wall, transformation of reality and illusion,
Floating layers dominate the role.
Focus on transformation of space,
That emerges, and is also invisible.

Cast/ Dressing Room and Bathroom of Main Bedroom
Vertical extension breeds the space of cross,
Bringing the lights straight along.
As a dim ink on plain paper,
There extension rolls.
Room of Light,
Return to the simple design of body and nature.

Cast/ Cabinet of Entrance
Between the mass,
Division of equal relations.
Scattering the concept of surface.
Rhythm of interest there it forms,
As a unique art of its own.

"世界是个大舞台，都有自己扮演的角色，都在演绎着属于自己的人生。" ——撷取自莎士比亚《皆大欢喜》。正如我们成就的空间角色，称职地扮演着。

楼梯：用一种线性连贯，链接复层场域，透过线性，影曳与建物间穿越流动。

客厅、电视墙：区分量体，素材堆栈延续；简朴却人文，随时间流动，光影在空间中静谧和谐。

茶室：旧桩交错形成，赋予结构强化空间量体。光线透过板材序列，平衡空间阴阳两式。长向机能延续坐落在端景的是一片宁静，以坐卧高度，开阔山景视野。古朴、低调、禅意缭绕，随遇而安。

光墙、车库、空桥：穿越光墙，虚实转化，层层悬浮，举足轻重。空间讲的是场域转换，显、隐，木石交迭，以不同的形式化分呈现。

主卧更衣室、主浴：纵向延续，衍生交汇空间。引光入室，枉直随形，如同白纸上点了一滴淡墨，延展而来。光室，回归人体与自然最为简洁的设计。

玄关造形柜：在量体之间，以相等关系划分，面的概念错落彼此，形成趣味的韵律，自成一道艺术。

DECORATIVE MATERIALS AND SPACE PLANNING | 装饰材料、空间规划

The designer continues his design feature, constructing the whole project with massive wood and stone, which are integrated, intersected and jointed in the whole space to present natural, quaint and humanistic visual effects. Linear field connections are coherent. Each function is independent but connected with each other; you will see scenery beyond the house and the house with scenery, creating a dialogue between each space.

设计师延续其一贯设计特色，以大量木质和石材构筑完成整个案例，彼此融合、交错、拼接呈现于各个空间，达到自然、古朴、人文的视觉效果。而线性的场域连结，流畅连贯。各功能区彼此分开又相互呼应，视外有景，景中有室，形成良好的空间与空间的对话。

NATURAL LIGHTING AND COLORS | 自然采光、色彩

The light is tranquil and harmonious with the flow of time. After penetrating the shutters and wooden fence, the light is brought into the rooms gradually and then drops into the deep well peacefully. A sense of quaint, low-key, adaptable and Chinese Zen spirit is prevalent in the whole space. The designer uses natural colors and lines skillfully, stressing naturism and adding some intimacy. The traditional culture is abstracted into simple and ethereal images. Little black element is applied, making the white color distilled and creating an interesting concise art.

　　随时间流动，光影在空间中静谧和谐。光线透过折叠的百叶窗、木栅栏，层层叠叠映照进室内，落在深井的是一片宁静。古朴、低调、禅意缭绕，随遇而安。巧妙地运用自然的色彩与线条，强调自然主义，有置身大自然的亲切感。同时将传统文化抽象成一个个简洁、空灵的意象置于室内。少量的黑色元素的注入，让空间的白得到升华，形成一道趣味简洁的艺术。

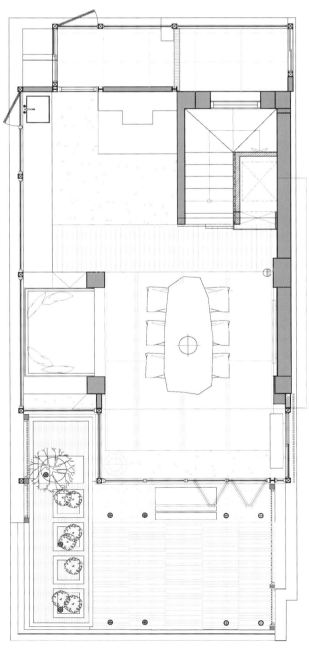

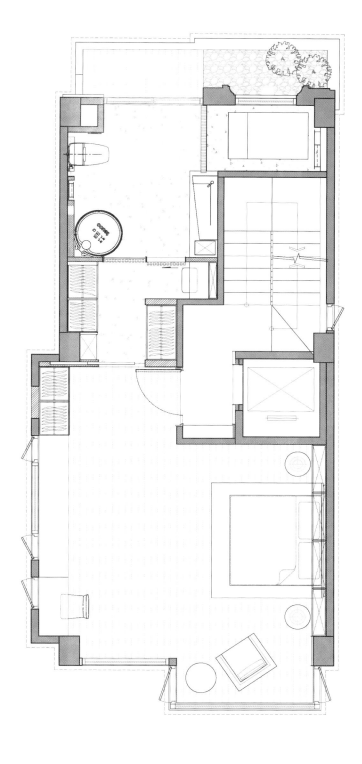

047

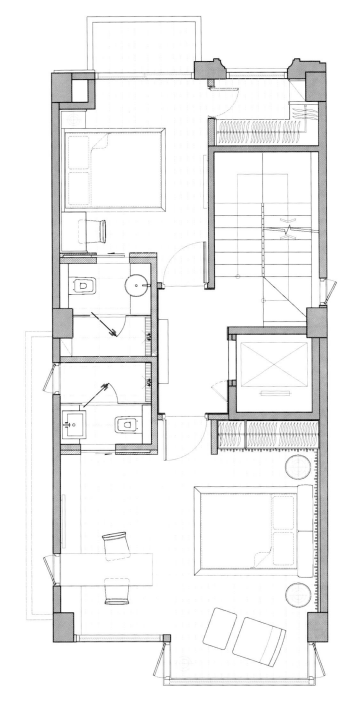

NATURAL STYLE | PRESENTING RUSTIC AND HUMANISTIC DIMENSIONAL DEPOSITS WITH MATERIALS CLOSE TO NATURE

THE AXIS OF LIFE

生活轴线

Location | Taipei, Taiwan
项目地点 | 台湾台北

Area | 230m²
项目面积 | 230m²

Designers | Tiangui Xu, Wenxin Lee
设 计 师 | 许天贵、李文心

Design company | TWA design
设计公司 | 传十室内设计有限公司

Photographer | Figure x Lee Kuo-Min Studio
摄 影 师 | 图起乘李国民影像事务所

Main materials | European imported tile, natural wood veneer board, latex paint, metal board, etc.
主要材料 | 欧洲进口瓷砖、天然木皮板、乳胶漆、金属板等

DESIGN CONCEPT | 设计理念

For an Eastern city, how to show the contemporary spirit and reveal the Eastern residential space is the focus of this project. In terms of style, it expects to embody the Eastern calm low-key artistic conception. But the four family members (the couple and two adult sons) have different stylish preferences for different spaces. So how to integrate different styles is a big challenge! Besides, the original surface of the wall of the building with an elevator is irregular, which is not good for planning the layout. Spaces redesign, axes reconstruction, spaces integration which is limited by the irregular layout and the requirements from the owners who want to have their exclusive bathrooms for three different bedrooms, and taking the mountain view into consideration make the designers divide the whole space into public area and private area. The public area is composed of two vertical axes. One is "entry axis", from hallway to dining room to recreational area, and the other is "landscape axis", from snack bar to dining room to living room.

对于一个东方的城市，如何展现具有当代精神却又能透露东方气息之住宅空间是本案切入之重点。在风格方面，主体上希望能体现东方低调沉稳之意境。但家人成员四人（夫妻与成年儿子2人），对于各空间之风格偏好都不同，如何汇整揉合多样风格是一大挑战！加上此电梯大楼之住宅单元，其原始外墙平面形状不规则，不利格局规划。空间如何重整，轴线再造与空间整合受限既有不规则平面格局条件，以及屋主对于三间卧室都希望有专属卫浴空间之要求，同时考虑既有之山景优势，初步将空间领域划分为公领域以及私领域。公领域以两条垂直轴线构成，其一为"入口轴线"，玄关－餐厅－休闲区，其二为"景观轴线"，轻食吧台－餐厅－客厅。

DECORATIVE MATERIALS | 装饰材料

Based on earth tones and rich-texture materials, it creates a warm low-key attitude. The walls of the axis-side view, such as TV wall, snack area wall and recreational area wall, are given contrastive textures and modeling, which indicates the theme of the space and the visual focus. Behind the simple and integral geometric forms are the collecting functions which the owners require.

以大地色系、肌理丰富之材料为基调，营造温润低调的内敛姿态，在轴线端景之墙面，电视墙、轻食区背墙、休憩区墙面等特别赋予对比性的材质与造型，带出空间的主题性与视觉焦点。简练而整体的几何形式背后，也隐藏了屋主要求的大量收纳机能。

COLORS | 色彩

Three bedrooms are designed upon the preferences of the owners. Bedroom A is based on clear water concrete supplemented by continuous folding ceilings and floors, which creates a natural and novel modern atmosphere. Bedroom B is based on black and white colors with the addition of avant-garde folded plate cutting technique, which creates fashionable tonality. According to the requirements of the clients to add some fashionable and low-key textures, the master bedroom has some wood veneer textures and metal textures between black and white. After redesigning the layout, the new spatial orders and axes relations are clear, bringing the landscape inside. The rich-texture materials of the earth tones show the outstanding attitude of the East and also create a new high-quality living standard for the owners.

三间卧室，各自以使用主人之偏好来构思。卧房 A 是以清水混凝土为基调，辅之以连续翻折的天花墙面与地板，营造朴质且新奇现代感之氛围。卧房 B 则是以黑白对比色系，加上折版切割的前卫造型手法，营造时尚调性。主卧室则应客户要求增加一些时尚与低调质感，在黑白之间增加了木皮色系材质与些许金属材质。经过格局重整，清晰地梳理出全新的空间秩序与轴线关系，将景观拉近室内。大地色系丰富肌理的材质，展现出东方大气之姿，也为屋主创造全新优质的居住质量。

NATURAL STYLE
PRESENTING RUSTIC AND HUMANISTIC DIMENSIONAL DEPOSITS WITH MATERIALS CLOSE TO NATURE

BEAUTY! CUSTOM-MADE HUMANITY SKYLINE

美哉！订制人文天际线

Project name ∣ Lee Mansion in Songren Road 项目名称∣松仁路李公馆	Location ∣ Taipei, Taiwan 项目地点∣台湾台北	Area ∣ 222m² 项目面积∣222m²	Project designer ∣ Haoqian Zheng 专案设计师∣郑皓谦
Designer ∣ Alan Yu 设 计 师∣俞佳宏	Soft decoration designer ∣ Weiting Zhong 软装设计师∣钟暐婷	Design company ∣ Shang Yih Interior Design Co., Ltd. 设计公司∣尚艺室内设计	Photographer ∣ Sam Cen 摄 影 师∣岑修贤

Main materials ∣ carved white, jade, stone husk, white oak steel brush wood veneer, fluoride acid gray lens, etc.
主要材料∣雕刻白、和氏璧、石皮、白橡木钢刷木皮、氟酸灰镜等

DESIGN CONCEPT ∣ 设计理念

The shadow falls on the log bench; the mountain is magnificent; is it a hallway? It is the designer's prospect for the space.

影滴落、原木长凳上泛起阵阵涟漪，衬以山石所述大气，是玄关，亦是设计师所给予的空间前景。

Move forward, among the scenery, the wood and stone resonate. Large amounts of white carved marbles symmetrically define the theme wall and stretch out. As for the above beams, the studio uses wood bending surfaces to outline the four-lever ceiling in the same slope, just like mountain ridges ups and downs, skillfully balancing the warmth of textures. The mountain ridges overlap level by lever and outline many trails. The carved stones present natural prototype. The wooden lines depict exquisite affects. The natural textures extend continuously. The natural elements matched with outside scenery spread to every space, sparkling into the pure textures through light and shadow.

　　步伐前行，景深间，木与石于天壁大胆共鸣，极具分量的雕刻白大理石，以对称之姿定义、延伸主题墙面存在，而天际在线的扰人梁体，工作室则以木质折面为笔，错以相同斜率、勾勒出四段层次的天花表情，大气层迭、铺叙犹如山棱起伏，巧妙平衡材质间的视觉温度。山的棱线层层迭错，勾勒出一道道的轨迹，石头的雕琢呈现自然的原型，木纹的线条刻划出细腻的感动，自然的肌理将绵延不绝的延续，以大自然的元素搭配窗外的景色铺陈于各个空间，透过光与影围绕洒落在纯粹材质中。

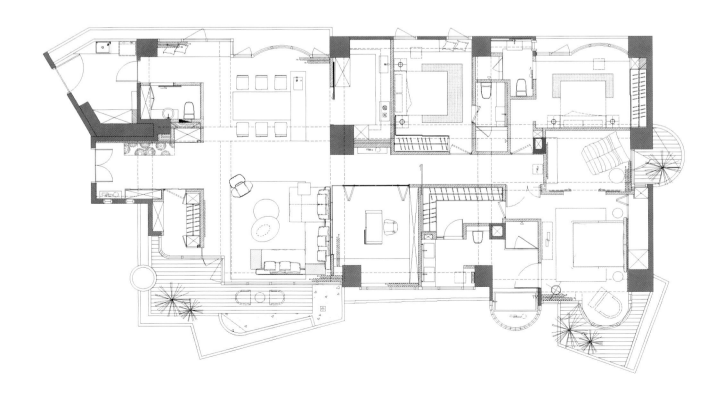

DECORATIVE MATERIALS AND SPACE PLANNING | 装饰材料、空间规划

Cold stones match with warm woods, which creates a conflict harmony. The natural textures warm and polish the neat lines. The concept of central axis connects all spatial functions and integrates the piecemeal space. From the hallway, natural wood husks are separated with irregular woods. Natural elements are used in the aesthetic spaces, such as symmetrical technique, stone with asymmetric patterns and harmonious division proportions. The symmetry of wall presents the balance of the middle axis.

Except for using wood and stone in the wall, the designers choose wood bricks in the public area from the ground to the wall to connect the large public area, and at the same time to skillfully get rid of the systematic impression of the connector and to redefine the possibility of high-quality residence.

  以冷冽的石材搭配温暖的木纹创造出冲突的和谐，自然的纹理温润了利落的线条，更以中轴的概念串连所有的空间机能，淡化空间的零碎性。自玄关起自然石皮与不规则的木纹分割，自然的语汇铺陈于美学的空间中，以对称的手法，不对花的石材，和谐的分割比例，利用墙体的对称性拉出中轴的平衡感。

  除以木、石作为天壁关系，公领域地坪部分，设计师则选用木纹砖为语汇，由地坪漫伸入壁面，串接起大公领域场景同时，也巧妙脱焦了接口的制式印象，重新定义精品宅邸的极致可能。

BRINGING SCENERY INTO HOUSE | 引景入室

The outside mountains stretch to the ceiling, high and low. The natural wood textures span over the end, which creates rich and restrained beauty of nature. The gray tone materials extend from the wall. The natural wood husk and the outside light follow with the neat lines, which creates silhouette of stretching mountains. The natural wood is the extension of kinetonema as if the branches shuttle back and forth in every corner of the house and grow naturally.

　　屋外山峰绵延持续延伸至天花板，高低交错，以自然的木头纹理跨越彼端，营造丰富与内敛自然之美。灰色调的地面材质延续自墙面，大自然的石皮和窗外的光线依循着利落的线条，创造出峰里绵延的剪影。自然的木纹作为动线的延伸，如同树的枝干穿梭在屋里每个角落，自然地生长着。

NATURAL STYLE
PRESENTING RUSTIC AND HUMANISTIC DIMENSIONAL DEPOSITS WITH MATERIALS CLOSE TO NATURE

THE JOURNEY OF TIME
时间旅程

Location | New Taipei City, Taiwan
项目地点 | 台湾新北

Area | 298m²
项目面积 | 298m²

Designers | Yu Pin-Chi, Tsai Yao-Mou
设 计 师 | 游滨绮、蔡曜牟

Design company | LUOVA Design Co., Ltd.
设计公司 | 创研俬集设计有限公司

Photographer | Jiahe Guo
摄 影 师 | 郭家和

Main materials | wood veneer, marble, iron rust board, kieselguhr, etc.
主要材料 | 木皮、大理石、锈铁板、硅藻泥等

DESIGN CONCEPT | 设计理念

The meaning of life lies in pursuing the essence of life, abandoning the fancy coat and levity to find one's value and the meaning of existence. The identity of space becomes more obvious; the richness of life makes the life journey colorful, which brings forth the individuality of space.

Art is the most important additive of practice and reflection in life. Where there is life, there is humanity; where there is humanity, there is art. Creating such a residence where art, humanity and life coexist harmoniously may call up the existence of higher life value.

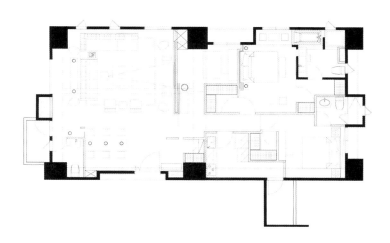

生活的意义是追求生命本质的存在感，褪去华丽的外衣，摒除浮夸后最终回到自我中心的价值，思索自我存在意义。空间主体的自明性会更加显著，借由生活上的丰富性，添加生命旅途过程中的色彩，空间场域个性由此萌生，而艺术是对生活的实践与印证最美味的添加剂。生活处处是人文，人文处处是艺术。打造一个艺术、人文、生活谐和的空间住所，期望由此案人们更能体会更高的生活价值意义的存在感。

DECORATIVE MATERIALS | 装饰材料

The fragrance and freshness provides a wonderful space atmosphere. Massive wood veneer is applied in the ceiling and wall, and the texture grows with the flow of time. The black and white marble TV wall is designed with good tension. The gray L-shape cloth sofa matches well with the daisy pattern and the round cushions, creating a feeling of vacationing.

　　自然的芬芳与清新的视觉，给了居住者良好的空间氛围感受。大量木皮的铺设，从天花到墙壁，一气呵成，清晰可见的纹理，是时间的生长。而黑白色造型大理石电视背景墙，张力十足。灰色的 L 型布艺沙发，配上盛开的菊花式样，加上旁边几个圆形的小布墩，浪漫休闲感仿若度假。

SPACE PLANNING AND NATURAL LIGHTING | 空间规划、自然采光

Other than the two-dimensional space planning, the design team is trying to find the viability of using 2.5D perspective principle to connect each space. Combining with the shadows on the wall created by the spectrum penetrated through the windows, each space is endowed with its unique function. In this way, the light in each space is maximized.

不同于二维空间图示方式布局空间区划，尝试利用 2.5D 的透视原理找寻各空间衔接处不同面向的可行性。结合光穿透建筑开窗面的线状光谱波长，映照在墙板之间所产生的阴影面，借此调配不同属性机能，由此让每个空间能获得最大采光面积。

NATURAL STYLE
PRESENTING RUSTIC AND HUMANISTIC DIMENSIONAL DEPOSITS WITH MATERIALS CLOSE TO NATURE

DIALOGUE AND FRAME
对话·框景

Location ǀ Taichung, Taiwan 项目地点ǀ台湾台中	Area ǀ 294m² 项目面积ǀ294m²	
Designer ǀ Yeh Chia Lung 设 计 师ǀ叶佳陇	Cooperating designer ǀ Jiaqing Ke 参与设计师ǀ柯家庆	Design company ǀ Leaves Architecture Interior Design 设计公司ǀ拾叶建筑+室内设计
Main materials ǀ paint, wood floor, tile, etc. 主要材料ǀ油漆、木地板、瓷砖等	Photographer ǀ Max Chung 摄 影 师ǀ钟崴至	

DESIGN CONCEPT ǀ 设计理念

Obscure boundary calmly and neatly defines the relationships among spaces and frames many beautiful family pictures. The spatial arrangement focuses on functional home with light, sound and people moving around, connecting the dialogue and relation between each other.
After a busy day, forgetting the business and taking a rest at home are no doubt the most precious things for people nowadays. A broader filed of vision, warm textures, the sensitive lights, the dearest family and the happy home all can make you get rid of the exhaustion of the day.

隐晦的边界，沉稳而利落地界定了空间之间的关系也框出了一幅幅家的美景。在空间配置上，以机能环绕家的核心为概念，光、声、人在空间里流窜移动，来往之间串联起空间彼此的对话与联系。

在繁忙中度过大半天，能暂时遗忘公事且回家好好休息，无疑是现代人每晚感到最珍贵的时光了！开阔的视野、温暖的木质、疗愈的灯光再加上亲爱的家人相伴，一同窝在共筑的幸福小窝，整个人的身心彷佛洗去了一天的疲惫。

DECORATIVE MATERIALS
装饰材料

The stainless steel frames vaguely define the location of living room and dining room. The study is divided by the transparent materials to maintain the independence and visual penetration. The wall of the bedroom uses warm textures and thin frames. The simple line sequences present moderate collocations. The warm afternoon sun lights sparkle over the ground and the wall. The simple textures emit warm smell and return to the original nature of the space through light and shadow.

空间中不锈钢边框隐约界定了客餐厅的位置；书房以透明材质作为分隔，保有各自独立性与视觉上的穿透；卧房的墙面使用温润的木纹及细框柜体，简单的线条序列，呈现温和的搭配。午后，温润的阳光渲染一墙一地，简单的材质衬托出温暖的气息，于光影之间回归原始的空间本质。

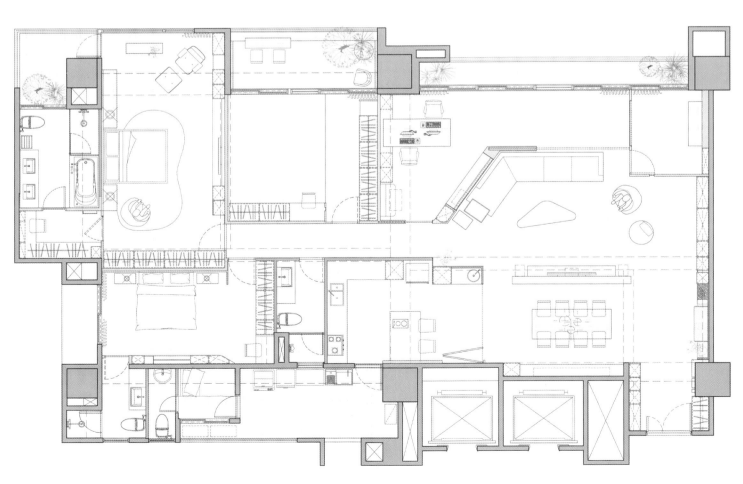

SPACE PLANNING | 空间规划

Regarding the living room as the center, there are dining room, kitchen, study and game area around. Although they are separate, they connect with each other and aren't limited by the life scales. The open vision makes the space more spacious. The scenery outside the window brings a little greenery. The space reflects the emotion. Thus there are scenes about families and dialogues between spaces.

以客厅为核心，边界环绕着餐厅、厨房、书房与游戏空间，虽各自一隅，却又紧密相连，不受局限的生活尺度，视野延伸使空间更通透。借窗景引绿，空间映情，来往之间，框出一幅幅家的景象，活络空间之间的对话。

NATURAL STYLE
PRESENTING RUSTIC AND HUMANISTIC DIMENSIONAL DEPOSITS WITH MATERIALS CLOSE TO NATURE

SENSITIVE RESIDENCE ON THE RIVER BANK WHERE WATER JOINS THE SKY
水天一色的河岸疗愈宅

Project name ｜ JJ House 项目名称｜淡水JJ HOUSE	Location ｜ Taipei, Taiwan 项目地点｜台湾台北	Area ｜ 181m² 项目面积｜181m²
Designer ｜ Tam 设 计 师｜谭淑静	Design company ｜ Herzu Design 设计公司｜禾筑设计	

Main materials ｜ oak floor, walnut solid wood veneer, tile, tawny glass, thin rock, stainless steel, titanium metal board, teak solid wood board, etc.
主要材料｜橡木地板、胡桃实木皮、瓷砖、茶镜、薄片板岩、不锈钢、镀钛金属板、柚木实木板等

DESIGN CONCEPT ｜ 设计理念

The couple left the foreign place where they had stayed for about 20 years because of another career and decided to move back to Taiwan to start their second half of life. Living for a long time abroad in a large space with different habits, the limited space planning is a big challenge to the designer. Based on the natural lighting, the beautiful scenery and the original structure, the designer brings natural elements into the interior by design techniques, making the kinetonema moving with the light, and adopts diverse and coordinated collocations of textures to make the space richer.

屋主夫妻选择离开待了二十多年的国外生活，为了人生的另一个事业，决定将生活重心搬回台湾，开始他们人生的下半场。长期住在国外大尺度的住家空间和生活习惯，在此案有限的空间规划上是一个很有挑战性的考验。同时在自然采光、风景优美与建筑物自身条件下，透过设计手法将自然元素带进室内，让动线沿着光线移动，用多样但协调的材质搭配，让居家空间的表情更为丰富。

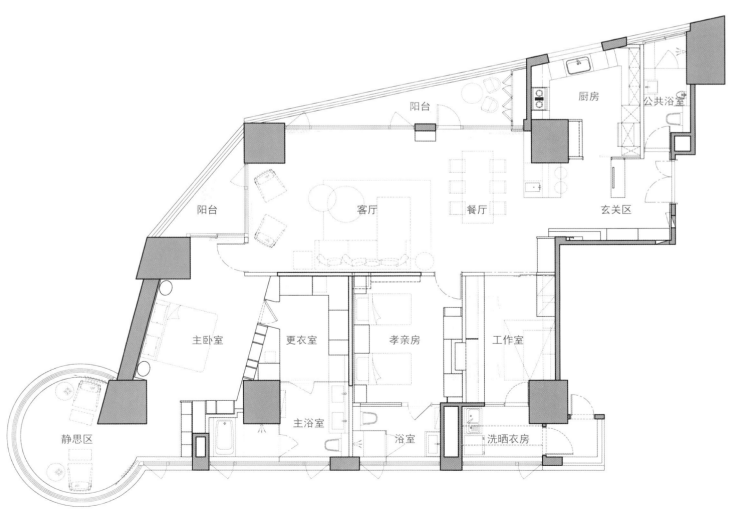

DECORATIVE MATERIALS ｜ 装饰材料

The hallway is in half-penetrated irregular degree, covered with titanium metal screen whose surfaces are treated in different way, and away from the wood chair. The asymmetric division of composite materials in the wall integrates the space by floor tiles. The principal axis of the study is the 4.5 meters' African teak solid wood table without cutting, with naturally sliced and jagged wood veneer covering the cabinet to echo the solid wood material. The elders' room uses earth color leather to cover the cabinet. The opaque glass meets the need of lighting and protects privacy in the bathroom.

玄关以半穿透的不规则角度且不同表面处理方式的镀钛金属屏风，脱开了悬空的穿鞋实木椅，壁面复合媒材的不对衬分割透过大地色系的灰色地砖来整合空间调性。以长达4.5无切割的非洲柚木实木桌板作为书房空间主轴，以自然刨切锯齿状的木皮作为柜体呼应实木材质。孝亲房以大地色皮革纹材质包覆柜体，透光不透影的玻璃材质满足了空间照明的需求也兼顾了卫浴空间的隐私。

SPACE PLANNING | 空间规划

Open the hidden door made by the iron-gray thin slate in the master bedroom, you can see that the wall is painted with special paint, matching with special irregular modeled wall lumps. This not only strengthens the visual endpoint extended from the living room, but also increases the depth of the entrance. Due to its structural conditions, such as non-squared wall and low ceiling, the master bedroom is redesigned. The wall above the head of the bed is painted with special paint. The opposite wall is decorated with large areas of bookcase made by the owner's favorite dentate gray wood veneers to make the declining space square. By the concept of declining roofs, the design of the ceiling meets the needs to modify the structure, pull up the height of ceiling and illuminate.

打开铁灰色薄片板岩的主卧隐藏门，壁面自然手刷纹的特殊涂料，搭配特殊配置不规则角度的造型壁灯，不仅使从客厅延伸的视觉端点加强，也让纵深较浅的入口空间得到转化。主卧室因建筑结构本身条件，非方正的墙面、低矮的天花等，透过设计手法让空间修饰完整。床头墙面以连续的自然手刷纹特殊涂料延伸，床尾壁面以对称呼应的角度，用业主喜爱的锯齿状灰色调木皮构成的大面积书柜，将原本歪斜角度的空间修饰成方正格局。而天花设计上透过斜屋顶的概念同时满足了钢梁斜撑的结构修饰、拉高天花空间和照明需求。

BRINGING SCENERY INTO HOUSE ｜ 引景入室

The living room and dining room have sufficient lights entering from the window. Outside the window, the natural river scenery, the blue sky and white cloud are the visual focuses that cannot be ignored in the space. Echoing the natural elements, the custom-made warm colored wood floors match with the special paint on the wall. Iron-gray thin slates of natural textures highlight the door as the focus of vision. The half-penetrated tawny yarned glasses and titanium metal frames make the single texture specially painted wall unvarnished and modest.

　　拥有大面积开窗采光的客餐厅空间，窗外自然河景、翠绿山景、蓝天白云是空间中无法忽略的视觉焦点。呼应自然景色的空间素材，以订制的暖色调木地板，搭配墙壁大面积的清水模特殊涂料；另外自然纹理强烈的铁灰色薄片板岩，将门片隐藏修饰成视觉端点的主角。墙面末端以半穿透的茶色夹纱玻璃和镀钛金属框让单一材质的特殊涂料壁面显得质朴而稳重。

NATURAL STYLE
PRESENTING RUSTIC AND HUMANISTIC DIMENSIONAL DEPOSITS WITH MATERIALS CLOSE TO NATURE

CUSTOM-MADE GRAND RESIDENCE

量身订制大气私宅

Project name ｜ Lin Residence in Banqiao Area 项目名称｜板桥林宅	Location ｜ Taipei, Taiwan 项目地点｜台湾台北	Area ｜ 265m² 项目面积｜265m²
Designer ｜ Tam 设 计 师｜谭淑静	Design company ｜ Herzu Design 设计公司｜禾筑设计	

Main materials ｜ marble, wood veneer, tile, glass, stoving varnish metal iron part, stainless steel, thin rock, wallpaper, etc.
主要材料｜大理石、木皮、瓷砖、玻璃、烤漆金属铁件、不锈钢金属、薄板岩、壁纸等

DESIGN CONCEPT ｜ 设计理念

With incredible space logic, the designer takes advantages of the original balcony to redesign the fixed layout and set the kinetonema, making the sunlight warm the whole residence. Every place manifests the designer's ingenuity and rigorous design concept and conveys her understanding of interior design. On the left of the hallway the marbles make the space grand. The glass screen with a layer of gauze inside tactfully removes the Fengshui taboo of crossing the house straightforward. At the same time she changes the kinetonema of the kitchen and sets a line-styled bar between the kitchen and the dining room to make the dining areas on a line and add a sense of level. Every space has a balcony with sufficient lighting, giving the family a wonderful life. As for the long gazebo outside the living room, two chairs and a telescope, you can have an exclusive observatory.

设计师发挥不可思议的空间逻辑，利用整体居家空间原有的阳台优势，翻转既定格局并切分居家动线，让和煦日光将整个复层空间温暖包覆。各处皆承载设计师的专属巧思与缜密的设计概念，传达设计师对于室内设计的自我真谛。宽阔玄关左侧以大理石为居家空间揭开沉稳大气的序幕，夹纱玻璃屏隔以不着痕迹的优美身段抹除穿堂煞的风水禁忌。同时更动厨房动线，在设有玻璃拉门的厨房与餐厅中间设置一字型吧台，使用餐场域连为一线，增添多元层次感。并且每一个空间外皆有采光完整的阳台，给予全家人通透美好的生活情调。而客厅外的长型露台，摆上两张单椅、架上望远镜，即可拥有专属天文台。

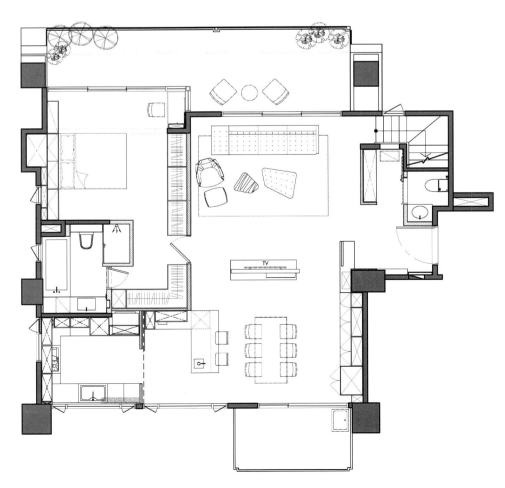

DECORATIVE MATERIALS ｜ 装饰材料

Because the family of five have the similar taste of style and tend to be low-key and calm, the designer makes efforts to choose materials and colors. Using wood veneers, marbles and tiles, with modern iron parts, stainless steel and yarned glass, the earth colors present a restrained and low-key style.

由于屋主一家五口对空间风格的喜好相近，倾向低调沉稳，因此设计师在选材与配色上特别用心。运用木皮、大理石、花砖等素材，佐以现代感的铁件、不锈钢及夹纱玻璃，以大地色系的协调色彩，传递内敛而不张扬的空间风格。

NATURAL LIGHTING | 自然采光

The designer turns the entrance of the guest bathroom next to the hallway to avoid its influence on lighting and hides the entrance by the marble wall with stainless steel frames in the living room. At the end of the wall, the designer creates a long window frame to bring in lights to make the hallway brighter. The open living room and dining room have large balconies. The designer adopts low furniture to highlight the advantages of lighting and ventilation and remove the shelter. The fresh color matches with the light textures, increasing the entire visual lightness and giving people a relaxing feeling.

设计师将玄关处的客用卫浴入口转向，避免影响采光，并以客厅中不锈钢框设的大理石墙面巧妙掩映，底部墙面则特别挖设长型窗框，引入楼梯间光线，让玄关不显阴暗。拥有大面积露台采光的开放式客餐厅空间，设计师采用低台度家具，目的为突显采光与通风优势，并抹去遮蔽感；色调清新以及清浅木质比例配置，更提高整体视觉明度，予人心旷神怡的轻松感。

INTERIOR VIRESCENCE
室内绿化

The decoration in the master bedroom is simple and fashionable. The fresh colored wood veneer wall is dotted with verdant plants which are vigorous under the sun light. In particular, the dangling bed is directly embedded in the wall as if it can make the resident have a floating dream.

主卧装潢简约不失时髦，色调清新的木皮背墙上，点缀着青翠可人的植栽，在阳光的照映下显得益发生机蓬勃。特别的是，悬空的床板直接镶嵌在墙面，彷佛能让居住者做出漂浮的美梦。

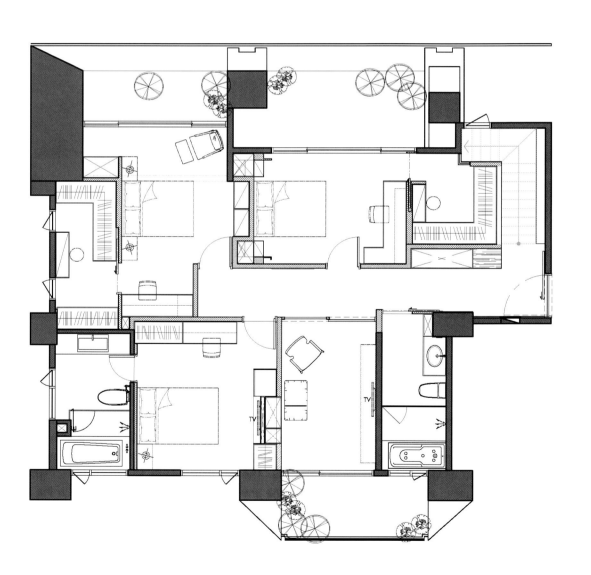

SPACE PLANNING | 空间规划

On life function planning, the designer not only readjusts the configuration in the kitchen, but also adds a bar for snacks and connects the kinetonema with the kitchen to be more fluent. The master bedroom and two daughters' rooms all have big dressing rooms and desks to meet their needs, and even have a long opening in the wall of the stairway to bring lights into the dark dressing rooms. Children's rooms have a multi-media room to give the children more private space. In addition, the big gazebo has comfortable outdoor furniture, which creates a relaxing space to enjoy the scenery.

在生活机能规划上，不仅将内厨房厨具配置重新调整，并增设吧台区，满足轻食需求，且让动线与餐厅连结，变得更流畅。主卧及两个女儿房均设置容量充足的更衣间，每个房间皆配置书桌，满足实用需求，甚至在楼梯间墙面开挖一个长开口，为阴暗的更衣室引光。而儿女房内加入的多元机能的视听室，给予儿女独立的利用空间。此外，偌大的露台，放置舒适的户外家具，形成一个悠闲赏景的休憩空间。

NATURAL STYLE

PRESENTING RUSTIC AND HUMANISTIC DIMENSIONAL DEPOSITS WITH MATERIALS CLOSE TO NATURE

A PLEASANT RESIDENCE WITH APPROPRIATE ARRANGEMENT

统整脉络·敞心居所

Project name ｜ A Villa in the Forest 项目名称｜林间·伪villa居所	Location ｜ Taipei, Taiwan 项目地点｜台湾台北	Area ｜ 248m² 项目面积｜248m²
Designer ｜ Hongyuan Chen 设 计 师｜陈弘芫	Design company ｜ He Zhi Interior Design 设计公司｜禾郅室内设计	

Main materials ｜ stone paint, aluminum grille, slate tile, slice stone, cultured stone, oak paint, iron part, glass, etc.
主要材料｜石头漆、铝格栅、板岩砖、薄石板材、文化石、橡木烤漆、铁件、玻璃等

DESIGN CONCEPT ｜ 设计理念

Taking in the simplicity and tranquility from Japanese culture to arrange the beautiful vision of the space, the designer views from the concept of "home", making some adjustments to present a new look for the dark house. In the overall planning, the designer integrates the beauty and function of the space and pays much attention to the details to interpret the pursuit of the family with four people.

撷取日式语汇中的质朴静谧，精心铺排空间的视觉之美，设计师以"家"为视角出发，透过格局调整与重新解构，为阴暗的老屋换上崭新面貌。在整体规划上，让空间美感与生活机能相互融合，详尽刻划内部的细节，演绎一家四口对于质感生活的诉求。

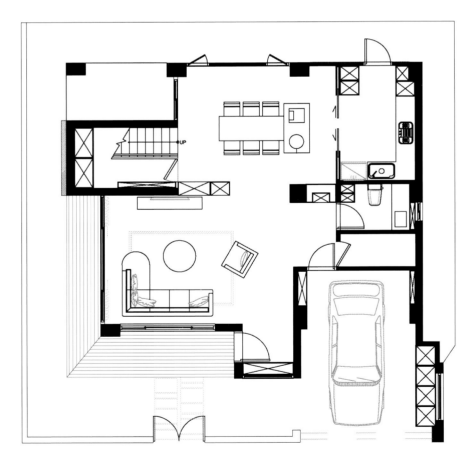

This case is a villa-style residence with three floors. Opening the door, you will come into the public area in the first floor where the complicated designs are replaced by the simple decorations in the highly opened space, providing a bright and clean atmosphere which makes for broad vision and pleasant mood. Natural and warm materials are used in the interior designs. For example, platane wood veneer is used for the cultured stone walls of facade, incase and rubble shapes to silhouette against the warm light. In this way, the texture and patterns and the changes of color show the spatial level, creating a pleasant sun villa.

本案为三层楼的别墅型居宅，推开门扉来到一楼公共领域，高度开放的空间以简约替代繁琐，呈现明亮而干净的清新质感，也让屋主一家视野开阔、心境更加澄净。设计师在场域内运用多种自然温暖的素材，例如使用梧桐木皮作为空间立面与柜面、乱石造型的文化石电视墙等，来映衬窗外的暖阳日光，借由纹路肌理与色彩变化，铺陈出空间层次，缔造亲切宜人的日光别墅。

DECORATIVE MATERIALS AND SPACE PLANNING | 装饰材料、空间规划

As for the architectural appearance, the grille mould presents contemporary Japanese style; slice stone is used for outwall decorations. The bamboo swaying in the breeze gently forms a cozy atmosphere. The design team makes each floor perform its function perfectly, adopting bright and clean design concept to provide private space for each family member. The master bedroom is decorated with simple lines to offer comfortable living experience. The main wall is decorated with clean wallpapers and ditch lines to shape the space with leisure and fashion. Entering into the lounge, you can see wood floor, wooden tables and desks as well as wooden showcase, which is not only comfortable, but also makes the overall Japanese style obvious.

建筑外观，以格栅造型围塑出现代感的日式语汇，并以薄石板材作为外墙的设计质材。户外的竹林植栽随风轻轻摇曳，形构绿意盎然的惬意氛围。禾郅室内设计将每层楼的场域机能，发挥得尽善尽美，以明亮洁净的设计主轴，构筑家族成员的专属空间。进入主卧房，运用简约、不复杂的线条，铺述整体性的舒适调性，采用干净的壁纸与沟缝线条为主墙，形塑休闲时尚的细腻品味。转进三楼的休闲室，全室木地板、木制桌椅及木作展示柜，既能感受温馨舒适的暖意，又可以增添整体性的日式风格与休闲语汇。

NATURAL LIGHTING AND BRINGING SCENERY INTO HOUSE | 自然采光、引景入室

The mirror style incase can enlarge the depth of the space, and the pending cabinet makes the dialogue between indoors and outdoors possible to bring a relaxed room. The L-shape double sided window brings adequate light into the living room on the first floor, creating a bright and spacious space. Meanwhile, cultured stone is used for TV wall, extending into the outdoor wall to make a free dialogue between people and window. Clean line processing together with the natural materials creates a soft and comfortable space. The third floor is a place for relax and communication, where the TV can be adjusted according to visitor's position. The beautiful dusk and magnificent sea cloud write a serenade in this space.

设计师以镜面式的柜面放大空间景深，悬空的柜体设置让室内外可以对话，让空间量体更加轻盈。L形的双面开窗，让一楼客厅可以盈满窗外的光线，营造明亮且敞朗的开放空间。同时以乱石造型的文化石来铺述电视墙，并由内延伸至户外墙面，引发人与窗外的自由对话。干净的线面处理，搭配自然的材质堆砌，淬炼出温婉舒适的空间概念。三楼空间是属于放松与交流情谊的场域，可做360度旋转的电视可以依照居者的观看位置调整。遥望美丽的黄昏与壮阔的云海景致，为空间谱出优美的小夜曲。

NATURAL STYLE
PRESENTING RUSTIC AND HUMANISTIC DIMENSIONAL DEPOSITS WITH MATERIALS CLOSE TO NATURE

A HOME WITH AXIS AND SCENERY

轴向·引景 家的聚场

Location | Taichung, Taiwan
项目地点 | 台湾台中

Area | 215m²
项目面积 | 215m²

Participating designers | Jiaqing Ke, Huiru Zhang
参与设计师 | 柯家庆、张惠茹

Designer | Yeh Chia Lung
设 计 师 | 叶佳陇

Design company | Leaves Architecture Interior Design
设计公司 | 拾叶建筑+室内设计

Photographer | Max Chung
摄 影 师 | 钟崴至

Main materials | oil paint, wood floor, tile, etc.
主要材料 | 油漆、木地板、瓷砖等

DESIGN CONCEPT | 设计理念

Located in the alley of the city, this villa is renovated from an old house. Just as other old houses, it has some shortcomings and it is poorly structured. However, the designers bring new life to the house. They bring new functions to the space, using vertical lines and horizontal ceiling and pavement to divide the areas, providing a simple but magnificent style. In this way, the openness of the space is maintained and each space is connected without interfering with each other, making the entire space more convenient. The arrangement of function and storage makes the space highly opened to satisfy the owner's expectation for home.

坐落都市巷弄中，翁郁围绕的透天别墅是由老屋重新整修而成，有着多数老屋皆有的壁癌问题与格局不佳之缺点，但在设计师的改造下褪去了旧屋印象，赋予居家全新的面貌。在空间规划上置入新的机能需求，因此以几道垂直墙体、水平天花与铺面界定各活动范围，打造简洁大气的居家风格。如此既保留空间的开放性，各个空间相互通连又不受干扰，提高空间连结使用上的便利性，并利用机能、收纳与空间上的整合，做到最大的开放感，充足的机能收纳，满足业主对家的期待。

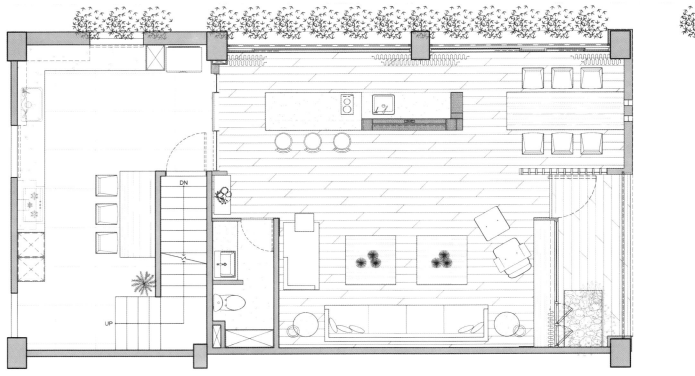

DECORATIVE MATERIALS | 装饰材料

Wood is widely used in its interior design, the cozy texture makes a comfortable atmosphere, and iron part is integrated, bringing out a calm tone and creating a harmonious space. There is no definite boundary in the interior design; axis connects with the space functions. Large-scale windows bring greenery into the house. The mild wood extends from the floor into the ceiling, bringing steady luster on the wall. Each space has its own scenery.

　　设计师大量地运用了木质，透过木纹的舒适质地营造出温馨氛围，并适当地融入铁件，让沉稳的色调注入屋内达到和谐。不刻意区分室内分界，以轴线连贯空间机能，大面积开窗平行于机能轴，将绿意映入室内，温润木色由地面延伸至天花，石材沉稳色泽垂直于墙面，举目纵观之下，空间自成一景。

SPACE PLANNING | 空间规划

To make a broad space, open style is applied in the public area, where long moving lines make the space interconnected without interfering with each other. Considering the park greenland is next to the residence, the sofa is faced to the window to make people enjoy the green scenery in leisure time. The long standing is close to the TV cabinet to divide the space. The dining area is close to the window, so that the family could enjoy the sunlight while working or dining.

为了使空间更显宽阔与大气，公共领域采开放式，长型动线让每个格局互通相连但不受打扰。由于住宅旁即是公园绿地，设计师刻意将沙发面向窗外，让人在休憩之余能饱览窗外的绿意风景。长型的中岛紧邻着电视柜，亦刚好作为与别区的间隔，餐区紧邻着窗边，让做家事或用餐的同时能享受日光沐浴。

BRINGING SCENERY INTO HOUSE | 引景入室

The road is next to the garden and the house is surrounded by green trees; all this bring abundant light and green shad into the house. The outer wall reflects the relation of the design elements. The decoration of bonsai and lamp creates a welcoming atmosphere, which becomes the most beautiful part of the space.

Dating and having a talk with your friends in your leisure time to relax yourself is a reflection of life. Making the best use of private spaces with favorable environment and thoughtful arrangement can provide a high-quality life.

相隔马路紧邻公园、被绿意所环绕的老房子，在大面开口下引入自然光线，大片的绿荫洒落室内各角落。同时屋内的关系显现于外墙上，错落之间，盆景与暖灯点缀，营造入口迎接之氛围，成为空间最美点景。

与友相约，开怀畅谈；闲暇时间，放松休憩，将生活的风貌完全体现。私域的空间运用环境良好对流与开放隔间的手法，连贯空间但保有独立性，增添惬意与舒适的生活质量。

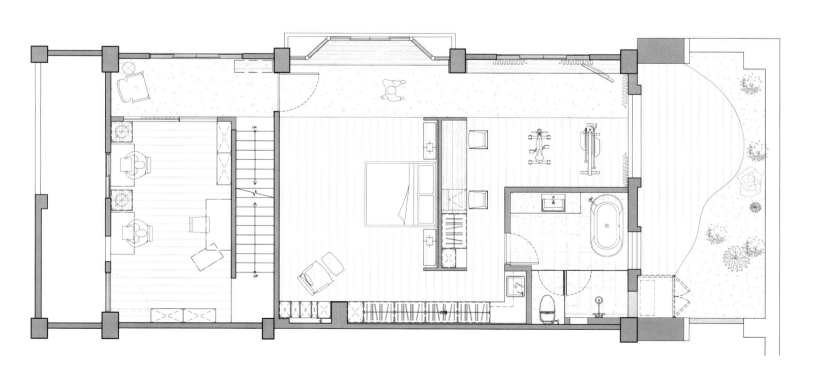

NATURAL STYLE — PRESENTING RUSTIC AND HUMANISTIC DIMENSIONAL DEPOSITS WITH MATERIALS CLOSE TO NATURE

LEISURE BLACK FOREST RESIDENCE

自然黑森林慵懒宅

Project name ｜ Zhao Mansion in Qing Chuan Fu Yu 项目名称 ｜ 青川富御赵公馆	Location ｜ Taipei, Taiwan 项目地点 ｜ 台湾台北	Area ｜ 179m² 项目面积 ｜ 179m²	Design company ｜ Shang Yih Interior Design Co., Ltd. 设计公司 ｜ 尚艺室内设计
Designer ｜ Alan Yu 设 计 师 ｜ 俞佳宏	Project designers ｜ Ruyu Lee, Yubin Hong 专案设计师 ｜ 李如愉、洪郁斌	Soft outfit stylist ｜ Weiting Zhong 软装设计师 ｜ 钟暐婷	Photographer ｜ Sam Cen 摄 影 师 ｜ 岑修贤

Main materials ｜ wood floor, iron part, transparent glass, black mirror, royal silver diamond, steel-brush wood veneer, stainless steel, grille, PANDOMO, etc.
主要材料 ｜ 木地板、铁件、清玻、黑镜、皇家银钻、钢刷木皮、不锈钢、格栅、盘多魔等

DESIGN CONCEPT ｜ 设计理念

The owner is a fashion designer who has a unique understanding of space planning. Massive special stone and stainless steel are used to build a mansion with favorable condition of lighting and ventilation, providing a natural leisure feeling like a Black Forest in Germany. The design team overcame the difficulty of carrying large size stones, making each stone perform its function and offering a space full of leisure nature atmosphere, which is most suitable for you to escape into the tranquility and stay in it with your cat leisurely.

从事服装设计业的屋主，对于居家空间规划有着独到的见解，使用大量特殊石材与不锈钢，为兼具采光、通风两者优势的大宅，打造满盈原始氛围的自然休闲感，令人仿佛置身德国黑森林。克服大量石材的搬运困难，将每一方石材各司其职，让整体居家空间仿佛从森林中提淬出来，散发悠然休闲的自然氛围，最适合慵懒地和爱猫一起窝在这方静谧中。

DECORATIVE MATERIALS | 装饰材料

The entrance is made from royal silver diamond, showing natural texture. The living room is decorated with stone husk-shaped TV wall, building a straight layering, creating a feeling of leisure. Missive lighting strikes a balance between the cool colors of the stone. Some bright decorations are placed on the balcony, which are in contrast with the green trees. The floors on the entire room are made of PANDOMO, extending into the sofa background wall, echoing with each other. Stainless steel materials are used in the bar to highlight its delicate texture. The ceiling in the dining room matches the white grille, showing the distinctive taste of the space. The master bedroom is decorated with wood floor and stainless steel wall, creating a comfortable sleeping area. Combining the function of study by the window, you will have a tranquil place in the sunlight. The bathroom is made from massive stones, which makes a comfortable living experience.

玄关使用的是皇家银钻石材呈现天然的材质纹理，而客厅则是利用石皮造型电视墙，打造富有况味的直向层次，营造全然休闲感，大面采光则平衡了石材的冷冽色调，更在外部阳台放置鲜亮单品，并增设木作与绿意，形成鲜明对比。全室地板材质使用盘多魔一路延伸至沙发背墙，材料相互串连呼应，吧台上使用细致不锈钢材质，突显精致的表层肌理。餐厅区上方天花搭配白色的格栅，线条式的划出空间的独特品味。主卧以木地板与不锈钢墙面相结合，打造罕见的睡眠空间，更在窗边结合书房功能，让阳光与木皮许你一个静谧小天地。卫浴以大量石材打造，让自然休闲感在全室蔓延。

COLORS | 色彩

The entire room is in a cold tune. Steel-brush wood veneer is used in the dining room, creating a cozy room full of warmth. The whole space focuses on the planning of soft decoration with a specially selected comfortable white sofa. Considering the owner has a cat, the materials of sofa's cloth can withstand cat's scratching. Enjoying your free time with a cat lazily will have a comfortable feeling.

室内空间偏冷色调，餐厅墙面结合暖材质钢刷木皮，用温暖包覆日常，更加适合人文居住。整体空间着重于软装的规划，特别挑选较慵懒舒适的白色沙发，因家中有养猫咪，所以布料的挑选上特别选用防猫抓的材质，喜欢和猫咪一起慵懒地坐在沙发上，让这股舒适感持续的在空间里发酵。

室内绿化
INTERIOR VIRESCENCE

自然采光
NATURAL LIGHTING

装饰材料
DECORATIVE MATERIALS

设计理念
DESIGN CONCEPT

色彩
COLORS

引景入室
BRINGING SCENERY INTO HOUSE

空间规划
SPACE PLANNING

LUXURIOUS STYLE

INTERPRETING HIGH-QUALITY LIFE IN A
CONCISE WAY

奢华风

以简约的设计手法诠释
奢华的高品质生活

LUXURY BARN LIVING

奢华谷仓生活

Location ｜ Taoyuan, Taiwan 项目地点｜台湾桃园	Area ｜ 780m² 项目面积｜780m²	
Designers ｜ Mac Huang, Carrie Meng, Eva Yuan 设 计 师｜黄士华、孟羿彭、袁筱媛	Design company ｜ XYIQ Interior Design Co., Ltd. 设计公司｜隐巷设计顾问有限公司	Photographer ｜ SAM 摄 影 师｜岑修贤

Main materials ｜ marbled floor tile, white paint, translucent rock plate, titanize stainless steel, fabric hard roll, etc.
主要材料｜石纹地砖、白色烤漆、透光岩石板、镀钛不锈钢、绷布硬包等

DESIGN CONCEPT ｜ 设计理念

This project inherites North American style, paying less attention to luxurious or complicated living to integrate senery and provide a high-quality lifestyle. The concept of luxury barn living cannot be simply concluded as roughness. Instead, it is about striking a balance among details. Stepping into the living room, you can see a sliding barn door, which is in sharp contrast with the outer wrapped organic shapes, thus bringing you to the BARN LIVING. The whole design conforms to the state of life; the sliding barn doors, black bricks and processed sycamore formed a distinctive style. The bar area is also geometrically shaped; the burlywood is in contrast with the old timber, providing a comfortable feeling. The bar has a dual function, serving for both parties and home use. The living room is a LOUNGE without a definate way for the arrangement of TV wall or furniture, which provides a causual feeling. The independent iron fireplace with firewood imported from Denmark brings back recollections of Vancouver life of the owner. The dark rock walls echo the texture of the old gray bricks. The design team take advantage of the gray bricks and sycamore and make the room without the feeling of renovation, blurring the boundary of old and new. The feather texture and gloss of the PANDOMO ground overloaded the cement.

建筑承袭北美风格，不过分强调住宅的豪华感或是复杂的造型，以融入景观并呈现质感的生活风格打造。自然原始生活的概念不仅是粗犷能概括而论，应该是在其中寻找细节的平衡。进入客厅前映入眼帘的是一扇谷仓拉门，与外侧包覆的几何造型形成强烈的冲击对比，从此进入了BARN LIVING，设计沿着生活状态进行，谷仓门、青砖与处理过的梧桐木形成强烈的风格。吧台区延续着几何造型作法，原木色与旧木头的对比，形成很舒适的生活感，吧台功能兼具朋友聚会与家庭使用。客厅为LOUNGE概念，没有明确的电视墙或是家具摆设方式，随性也随意的感受。从丹麦设计生产的着铁材火独立壁炉，乘载着屋主在温哥华生活的记忆。深色岩石墙面呼应旧青砖的肌理与质感，我们借由青砖刻意不填缝的作法与梧桐木的处理，让空间没有新装修的感觉，将时间的痕迹停留，分不清新与旧，PANODOMO地面的羽毛纹理与光泽，使水泥产生了层次与感觉。

DECORATIVE MATERIALS | 装饰材料

The master bedroom carries the design concept of the living room; the sleeping area belongs to the rest area. There is no excessive shape or light disturbances in the bedroom. Cushions on the sturdy headboard make the master bedroom cozy together with natural wood and solid hardwood floor, enriched by leather paint and stucco. The dressing room is made from genuine leather and plated titanize stainless steel, deliberately lifting the ground floor to avoid the moisture. The bathroom is another focus of living space; the spacious bathroom includes a steam shower, a bathtub and a washbasin for couples. The details are the main axis of the space. The diamond-shaped bathtubs take the storage of temperature and water drainage system into consideration. The washbasin is decorated with a 5mm brushed stainless steel and concrete bricks. The study also functions as a drum room, integrating rock spirit. The designers use metal to match the genuine leather and dark colored space while taking the acoustic insulation, acoustic absorption and noise reduction functions into account.

　　主卧室承接着客厅的想法，睡觉是休息的区域，没有过多的造型与光线干扰，足够份量的床头染色牛皮软包造型，稳定了睡眠质量，两侧的原木使空间充满人性的温润感，羊皮革漆的天花与水泥状的马来漆在粗犷中增添细节。更衣间由真皮与镀钛不锈钢打造，刻意抬高地面避免湿气。主卧浴室是另一个生活重心空间，偌大的空间包含了双人淋浴间兼具蒸气室、双人浴缸与双人脸盆，粗犷中的细节是空间中的主轴，钻石型浴缸考虑了温度保存与快速排水机制，洗手台使用5毫米拉丝不锈钢与墙面混泥土砖搭配。书房兼具练鼓房功能，融入摇滚概念，金属搭配真皮与深色系的空间，并考虑了隔音、吸音、降噪的功能。

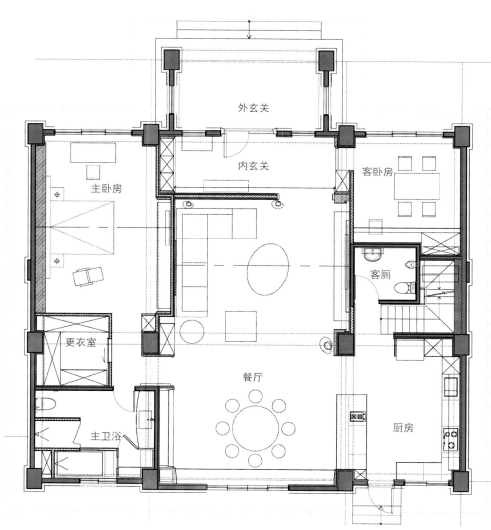

SPACE PLANNING | 空间规划

The entire space extends its architecture style with a feeling of simple American living. The spacious entrance wall protects the privacy of the owner, and provides many art crafts for visitors to appreciate. In the middle of the open dining room, there are some tea tables. The designers take life functionary as the main axis and extend their design concepts. The dining room is a place for family gatherings. The master bedroom emphasizes peaceful and cozy feeling. The main bathroom provides SPA functions, with spacious space making for comfortable living. The staircase acts as a transaction space for the conversion of life patterns.

空间延续建筑的风格，美式的生活居家感，偌大的玄关墙保留住户的隐私空间，以艺术品欢迎朋友的到访。开放式的客餐厅中间摆设茶桌，以生活机能为主轴扩散设计思路，餐桌与中岛是家人经常聚会的地方。主卧室强调宁静安逸，主浴室空间置入 SPA 功能，较大的空间让生活更舒适。楼梯为转换生活模式过渡的空间。

COLORS | 色彩

Strong colored American simple style with warm colored room and bathroom provide a comfortable living atmosphere. The dark purple is in contrast with the white color in the living room; the dyed leather soft roll on the bed adds some romantic felling; the independent fantastic white dressing room and the distinctive bathroom add some vividness.

强烈色彩的美式简奢风格，暖色系的房间与卫浴空间强调生活氛围。客房内深紫色对比白色，床头与床尾的染色真皮软包增加空间中的浪漫感，白色梦幻独立更衣间与强烈设计感浴室，打造生活感。

LUXURIOUS STYLE | INTERPRETING HIGH-QUALITY LIFE IN A CONCISE WAY

THE BEAUTIFUL RURAL RESIDENCE

乡林景居

Location | Taipei, Taiwan
项目地点 | 台湾台北

Area | 297m²
项目面积 | 297m²

Designer | Kai Zhang
设 计 师 | 张凯

Design company | THE Elegant Interior Design
设计公司 | 惹雅国际设计

Main materials | black mirror, metal, black iron, titanium, spray lacquer, wood floor, fabric, gray stone, etc.
主要材料 | 黑镜、金属、黑铁、钛金属、喷漆、木地板、绷布、城堡灰石等

DESIGN CONCEPT | 设计理念

Materials, configurations, lines, colors, shape proportion and formation are chosen carefully in this project. The designer begins to look for a new atmosphere, and it could be a Zen-like life or natural aesthetics. Flipping away the dust and noise and returning to the earthy mood, the mountain view is softly brought into the living spaces. Deliberately simplified layout leads blundering minds to deposit and quiet. The wall is in the original state with towering rough stones, propping up the tall roof of the structure.

本案每一个空间的材料、配置、线条、色相搭配、形体比例与构成都经过非常缜密的琢磨与筛选，设计中我们开始寻找一种新的气场、新的氛围，它可以是一种生活禅意，也可以是一种自然美学。拂去尘嚣喧扰，回归朴实心境，山林景致悄悄引入居所空间。刻意简化的铺陈，引导浮动的心灵慢慢地沉淀与静逸，将墙面表情回归到材料的初始状态，笔直高耸的粗犷石材山壁，俨然撑起建筑物体挑高的盘顶。

SPACE PLANNING | 空间规划

The public space focuses on log stacks in the dining room as if piling up the family's life from different life experiences of the family members. The owners can collect what they like in this space. They desire to enlarge the public space, merging the open kitchen with the rest area in the living room to become a complete public space.

让公共空间的一景，聚焦于餐饮空间的原木堆栈，如同堆砌家人的生活片刻，来自家庭成员不同生活经验的汇集。可以恣意地收藏于此交叉轴线空间，极度的欲求扩张公共生活区域，将餐饮开放式厨房区与客厅休憩空间合并，串连成完整的公共区域。

NATURAL LIGHTING | 自然采光

The whole space is in storied structure. Every private space is separated and brings large scales of lighting into the interiors. The outdoor scenery echoes the indoor tranquility, performing a beautiful view between dark and bright. Living here, you can enjoy the outside scenery at home in a leisure afternoon, or go to nature to feel the comfort.

整体空间上下错层，各部位私领域互相区隔，每个区域都采引大范围的窗外光景，引进室内空间。户外光景与室内沉静相互鸣奏，相应和声，明暗之间演绎一室居家好风景。居住于此，午后闲暇均可邀景入室，也可倾心融入自然中去，清闲淡然处之。

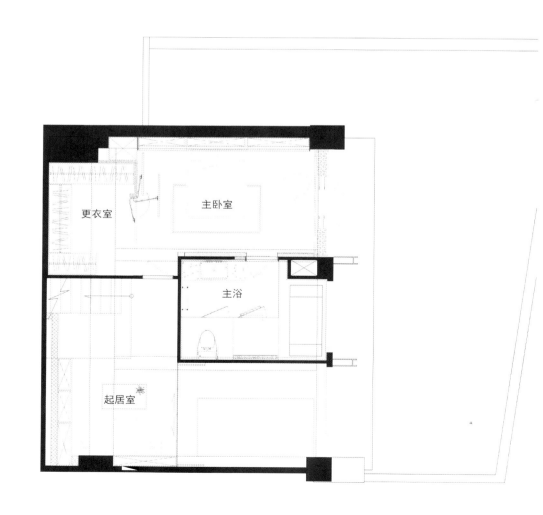

LUXURIOUS STYLE
INTERPRETING HIGH-QUALITY LIFE IN A CONCISE WAY

IDEALISM AND ARBITRARINESS

唯心·随意

Area ｜ 350m²
项目面积｜350m²

Designer ｜ Luke Wang C.H.
设 计 师｜王俊宏

Design company ｜ W.C.H Create Art Interior Design
设计公司｜王俊宏室内装修设计

Photographer ｜ Kyle Yu
摄 影 师｜KPS游宏祥

Main materials ｜ marble, wood floor, wallpaper, iron part, etc.
主要材料｜大理石、木地板、壁纸、铁件等

DESIGN CONCEPT ｜ 设计理念

Stepping downstairs, hidden in the noise of the city, stones, greenbelts and fresh garden scenery come into your eyes. You can have a cup of tea or chat with your friends until night falls. Entering the room, you can see wonderful pictures. Upstairs, the fragrance sprays into your face as if tracing back to the ancient times, making you feel warm and comfortable. The house is in stored structure according to the up-and-down slopes. When morning draws, light and shadow wander in the rooms. Downstairs, you enter the house. The broad yard is covered with stones and greenbelts to keep away from the noisy city. In this house, you can enjoy the tea and have a talk without interference.

踏阶而下，隐身于市井喧嚣中，飞石、绿地，清新庭园景致映入眼帘。或尽兴论茶、或秉烛夜谈直至夜幕低垂。入室乍见缤纷画意，方圆之间，巧妙互现。当东方既白，光影游移于厅堂，拾级而上，馨香暗自漂浮，彷佛穿越古今，暖意上心，随遇而安。随坡地起伏而建的复层格局，踏阶而下，即进入建筑基地，以飞石、绿地铺陈延伸，让开阔的庭园，隔绝世俗喧嚷的尘嚣，置身其中，得以安心自在品茗论茶、秉烛夜谈，不受干扰。

DECORATIVE MATERIALS | 装饰材料

Passing through the yard and entering indoor, on one side there is a square living room, and on the other side there is a small round dining table. The design of square and round is like the announcement of the frames with paintings from Murakami Takashi on the wall. The Chinese and Western furniture match with each other perfectly, bringing a quaint feeling into the modern neat lines.

穿越庭院进入室内，一边是规矩方正的客厅，一边是圆融小巧的餐桌，方、圆之间的造型、尺度拿捏，正如墙上村上隆缤纷画作框架的宣示。中西合璧的家具完美融入，在现代感的利落线条中，纳入一丝古意。

COLORS | 色彩

The color of the window frames is calm and restrained, outlining the lines of the space. Deliberately decorated with warm wallpapers, the warm tones create a comfortable living atmosphere. Stepping upstairs and entering the private areas, the Eastern Zen-like wood textures render elegant and cultivated artistic conceptions.

沉稳内敛的窗框色彩，勾勒出空间线条，刻意搭配略带暖意的壁纸，柔化阳刚冷调的用色，营造居家温馨气氛。拾级而上进入私领域，延续公共空间充满东方禅意的木质调用色，渲染书香淡墨痕的文雅意境。

GO HOME OR TRAVEL

回家·旅行

Project name ｜ Up Town 项目名称｜UP TOWN	Location ｜ Taipei, Taiwan 项目地点｜台湾台北	Area ｜ 150m² 项目面积｜150m²
Designer ｜ Sean 设 计 师｜张祥镐	Design company ｜ Etai.Space Design Office 设计公司｜伊太空间设计	Photographer ｜ Kyle Yu Photo Studio 摄 影 师｜游宏祥摄影团队

Main materials ｜ stone, titanium metal board, wood veneer, iron part, solid wood floor, slice stone, stoving varnish glass, stainless steel, etc.
主要材料｜石材、钛金属板、木皮、铁件、实木地板、薄片石材、烤漆玻璃、不锈钢等

DESIGN CONCEPT ｜ 设计理念

A couple who loves travel met a designer who was traveling in New York. Under different time and space, the designer created a rudiment of the "home" inspired by the observations along with the journey.

爱旅行的一对夫妻，遇上了正在纽约旅程上的设计师，不同时空下激荡出创作火花，将旅行中的观察内化成灵感的来源，完成一个"家"的雏形。

173

"Home" is the beginning and the end of a day, where life memory communicates with emotion. In order to shape a warm cohesion for the family, the designer skillfully blends in the traveling atmosphere, bringing in different cultures from all parts to enrich the family's life. The design of this project gives the owners a special feeling at home as if they are traveling. The designer uses composite materials to create a perceptual and rational living style.

"家"是生活一天的起点与终点，生活记忆与情感交流的地方，为塑造家庭温聚向心力，巧妙地融入屋主爱旅游的氛围，从各地风情文化作为丰富家与生活涵养，本案给屋主一种回到家也能感觉在旅行的心情。结合设计师善用复合媒材的设计手法，呈现空间质感，打造一种感性又不失理性居家风格。

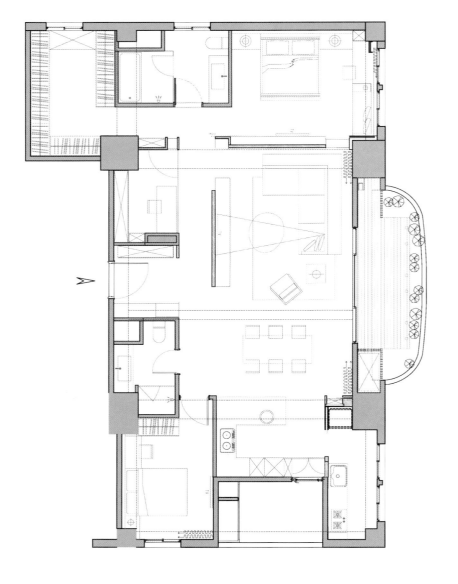

DECORATIVE MATERIALS ｜ 装饰材料

In the hallway, it continues with composite materials by using iron part to support the collecting cabinet to create a sense of impending, which shows the aesthetics of structural design. The design adopts composite materials to create vivid visual effects and a sense of stratification, blending in aesthetics of arts and crafts with the life tastes. The originality from different views and travel experience presents a design with concrete composite materials and a beautiful artistic conception from humanity and soul.

从玄关延续复合媒材运用，以铁件支撑收纳柜体营造悬空感，结构美学的妙用在此表露无遗。设计师以复合媒材创造活泼的视觉效果以及空间层次感，将工艺美学融入生活品味，从不同视野与旅行经验撞击出创意火花，展现一项具象的实体复合媒材设计之作，更是人性与心灵交流出来的一个美好意境。

SPACE PLANNING | 空间规划

Living space shows human personality. Through the communication between experience and aesthetics, they approve each other so as to solve problems of design details. The designer indicates that the feature of this project locates in the place of the TV wall. Besides flexible and leaping sense of stratification, it can look out from the bottom to the outside 101 scenery in Taipei, rather than glimpse everything with one look. The living room presents a balance between dynamic and static and a low-key luxurious humanity. The decorations of furniture and sofa integrate with the entire design to shape the home aesthetics. In the ratio of space and the layout of kinetonema, it gets rid of the original frame to create the flexibility of the space naturally. To decorate not for decoration presents coordination. Long aisles of living room, dining room and kitchen and the ceiling in strip shape, enlarge the breadth of the space and create a unique style.

生活空间体现"人"个性，透过经验与美学的交流，彼此认同，便能解决居家设计的细节问题。设计师表示本案的独特点是电视墙位置，除富有灵活与跳跃层次感，避免一眼望穿的视觉感受，能从底道远望窗外台北101景观。客厅展现一种静动态平衡、低调奢华饶富人文气息，家具沙发等装饰取材，适切地融入整体设计，形塑一种家的美学。在空间配比上与动线布局，跳脱原有框架，自然地营造空间灵活度，不为了装饰而装饰，展现协调性。在客餐厨长向走道，以条状式的天花板，延伸空间广度，勾勒独有气派感。

With regard to the privacy and a faint mystery, the wall of the bedroom is designed in a groove shape. From the corner of the living room, you can see the bedroom subtly. The independent sliding door protects complete privacy. The master bedroom displays a hotel-style delicacy, using different fabrics, iron part and wood veneer instead of paint to meet the living particularity of the couple who loves travel. Activating the life experience offers not only a relaxing place but also a lifestyle. As if the purpose of travel is to enrich the connotation of life and provide a warm and tolerable corner for soul. Home is sublimated from a building constructed of reinforced concrete to a residence full of happiness.

　　为顾及私密又能营造隐约神秘感，主卧房面墙凹槽设计，从客厅一角便能隐约看到卧榻，独立拉门设计又能兼顾完全隐私的遮蔽使用。主卧室设计展现饭店精致化，运用不同媒材绷皮、绷布、铁件与木头贴皮取代墙刷油漆，满足爱旅行的夫妻们对生活品味的讲究度。活化家的生活经验，不只是提供休息睡眠的场所，而是一种生活方式，就如同旅行的目的，充实生活与生命的内涵，提供心灵一个温暖宽容的角落。家，从钢筋水泥打造的空壳，升华成满载幸福的住宅空间。

LUXURIOUS STYLE
INTERPRETING HIGH-QUALITY LIFE IN A CONCISE WAY

TRANQUILITY
静谧

Location ∣ Kaohsiung, Taiwan 项目地点 ∣ 台湾高雄	**Area** ∣ 153m² 项目面积 ∣ 153m²	
Designer ∣ Chris Wu 设 计 师 ∣ 吴冠谚	**Design company** ∣ Cj Interior Design Co., Ltd. 设计公司 ∣ 长景国际有限公司	**Photographer** ∣ Ar-Her Kuo 摄 影 师 ∣ AR-HER KUO
Main materials ∣ iron part, wood veneer, stone, etc. 主要材料 ∣ 铁件、木皮、石材等		

DESIGN CONCEPT ∣ 设计理念

Tranquility is a state of conciseness and is a balance not going by with time. After fast and short sensory stimuli, it still remains charming cultural connotations. Without exaggerated decorations and abrupt designs, you can only feel the relaxing, stress-free and comfortable living atmosphere.

静谧是一种心境的简约，是一种不会随着时间而流逝的平衡，在跳脱虚妄或短暂的感官刺激后，依然保有耐人寻味的人文深度。没有浮夸的装饰符号，也没有争芳斗艳的突兀设计，这里的方寸之间带给入室者一种放松、零压、舒缓的居家氛围。

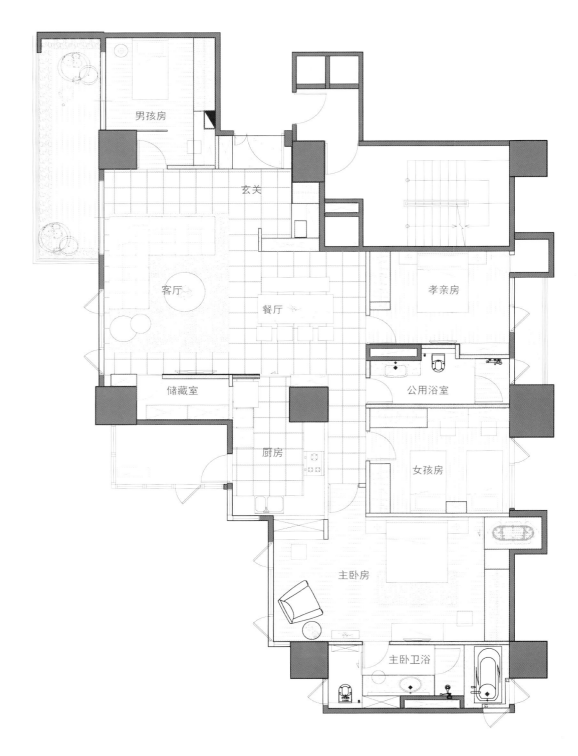

DECORATIVE MATERIALS | 装饰材料

The marble of the hallway is the prelude to the space. The floor uses the same tiles with different tactile impressions. The permutation and combination of different proportional geometric figures show multi-expressions of the space. The entire space combines exquisite stones with warm wood veneers, with soft screen window bringing in the outside lights, which forms a tranquil and harmonious atmosphere. The bedroom is covered with wood floors continuing the tone in public area. The jumped-color soft decorations offer some fun to the tranquility. The reading area is next to the window, which offers the readers the most comfortable lights.

玄关的大理石为空间揭开大气的序幕。地面采用同款瓷砖搭配三种不同触感，以几何比例分割排列组合，展现出空间的多重表情；整体空间以细致的石材搭配温润木皮，并于一旁配置柔纱帘，引进户外的温柔光影，形成静谧与和谐的氛围。卧房空间则铺陈木地板，延续公领域之调性；佐以跳色软饰，在静谧中带出一点玩味。并将阅读空间分配于靠窗处，让居者感受最舒适温柔的采光。

COLORS | 色彩

The quiet gentle tone warms and amazes time, leaving a sweet smile. A brush of bright yellow adds a sense of stratification to the space. The round black leather tea table looks calm. The warm yellow wood dining table and chairs make the dining time more leisure and casual.

悄无声息的柔和调性温柔了岁月，也惊艳了时光，盈一眸恬淡，随遇而安。客厅的一抹亮黄为米色空间增添了层次感，黑色的皮质圆几沉稳了空间。暖黄的木色餐桌椅让用餐时间更加悠闲随性。

LUXURIOUS STYLE
INTERPRETING HIGH-QUALITY LIFE IN A CONCISE WAY

ENJOY LUXURY
尊享奢华

Project name ｜ Bai Da Fu Li 32A Showhouse in Taichung 项目名称｜台中百达馥丽32A	Location ｜ Taichung, Taiwan 项目地点｜台湾台中	Area ｜ 799m² 项目面积｜799m²
Designer ｜ Tzu-Yeh Yang 设 计 师｜杨自晔	Design company ｜ Original Design Company 设计公司｜原创美学设计	

Main materials ｜ stone mosaic medallion, metal, shell, wood, glass, wood veneer, wood floor, etc.
主要材料｜石材拼花、金属、贝壳、木作、玻璃、木皮、木地板等

DESIGN CONCEPT ｜ 设计理念

This case focuses on providing excellent reception and providing presidential suite experience with hotel style design concept to make an honorable life. Stepping into the entrance, you will see the calm expression of space decorated with black olive marble floor; its momentum successfully attracted the attention of the guests. In the living room area, the designer uses inlay style shaped line and double L-shape sofa, removing the image of gallery skillfully to create a broad public area. A glass door is used to divide the living room and dining room which keeps the space outspread.

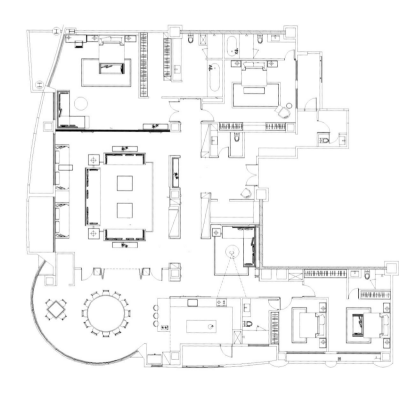

以精品招待会所为主轴定位，透过饭店式的设计概念，创造媲美总统套房等级的格局机能，让生活成为至高无上的尊荣享受。步入玄关，以黑橄榄大理石铺陈空间的沉稳表情，其不凡气势成功吸引了宾客的目光；来到客厅区域，设计师以回字型动线与双L型的沙发配置，巧妙地将廊道意象破除，创造尺度开阔的公共活动空间，并以一道玻璃拉门作为客、餐厅的区域界定，让空间感持续延伸。

SPACE PLANNING | 空间规划

The designer uses suitable color, light and texture to arrange a small sitting room in the aisle which reflects owner's taste. The double L-shape sofas in the living room provide a spatial interactive room. The carpet is replaced by the stone mosaic medallion, which not only divides the areas, but also is convenient to clean. The delicate crystal lamps represent the pulsation of wind, bringing magnificent and aesthetic visual effect.

设计师于过道空间规划小起居室，色彩、灯光与异材质的运用恰到好处，充分展现居住者的生活品位。客厅区域双 L 型的沙发配置，提供宾客宽敞舒适的互动空间。地面设计以双色石材拼贴手法取代地毯的铺设，不仅作为场域界定，同时也兼顾清洁的方便性。利用精致华丽的水晶灯具，其流线造型象征风的脉动，为空间带来既大气又唯美的视觉感受。

COLORS | 色彩

The master bedroom is decorated with bright and quietly elegant color, and the free space is equipped with sofas to make the living experience of presidential suite. The second bedroom is decorated with lavender bedside facade and exquisite table lamps to create an elegant and romantic space.

　　主卧室以明亮淡雅的色调铺陈其个性表情，运用空间余裕设置沙发区，打造总统套房等级的生活享受。次卧室紫藕色系的床头立面，搭配床侧精致的造型桌灯，表现出浪漫典雅的空间调性。

LUXURIOUS STYLE INTERPRETING HIGH-QUALITY LIFE IN A CONCISE WAY

A GORGEOUS NEW YORK STYLE RESIDENCE
华丽的纽约

Location ｜ Taiwan 项目地点 ｜ 台湾	Area ｜ 156m² 项目面积 ｜ 156m²	
Designer ｜ IDAN 设 计 师 ｜ 江欣宜	Design company ｜ L'atelier Fantasia 设计公司 ｜ 缤纷设计	Photographer ｜ Kevin Wu 摄 影 师 ｜ 吴启民
Main materials ｜ black herringbone tile, stone, crystal, gold foil, etc. 主要材料 ｜ 黑色人字拼瓷砖、石材、水晶、金箔等		

DESIGN CONCEPT ｜ 设计理念

According to the owner's demand, the designer transformed four rooms into three rooms and readjusted refined materials and craft elements. The minute way gives the house a new look and transformed into a celebrity home. The entrance is decorated with layered ceiling together with luxurious spheroidal crystal chandelier to present elegant light. The artistic paintings on the wall and the unique sofa reflect the owner's elegant taste. Moving forward into the house, you will come to the living room where the texture of the white marble extends through the whole TV wall, which becomes a magnificent yet delicate part of the space. The wall lamps on both sides are Portuguese fine crafts, which add some elegance and warmth.

设计师依据居住者的生活需求，将4房改为3房，并重新调整格局配置比例，在空间中注入精致材质以及工艺品元素，细腻的手法表现，让毛胚屋脱胎换骨，变身为具有绝佳品味的质感名流宅。玄关处天花向上做出层次，奢华璀璨的球型水晶吊灯从中垂坠，在天花深处形成雅致光影，墙面挂上艺术家画作与造型别致的沙发，在进门的一开始，即暗示屋主的优雅品味。继续向内进入客厅，阿拉伯白大理石的天然脉纹蔓延整面电视墙，成为壮丽而细腻的空间端景，两旁设置的壁灯则是来自葡萄牙的精美手工艺品，典雅温馨。

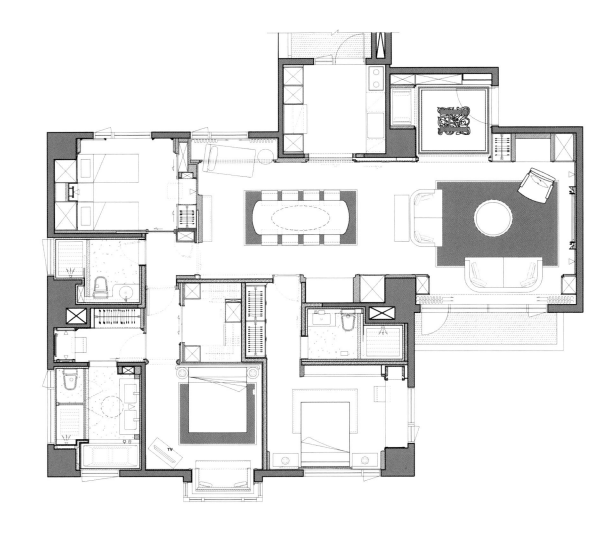

DECORATIVE MATERIALS | 装饰材料

The dining room is elaborately built with thoughtful arrangement of the details. The luxury lighting creates party atmosphere, and artistic works are also placed. Even the kitchen door plates represent hundred years of fine craftsmanship, and manual gold foil frescoes are presented with plant patterns. The stunning but exquisite art crafts become the best topic while having a feast. The wall, washbasin and the bathtub in the bathroom are made of natural white marble, and the arch-shape ceiling brings a sense of European court into the room. A chandelier purchased from Florence makes for a noble atmosphere while taking a shower.

　　精心打造的用餐空间，每个细节都相当讲究，餐桌上方以奢华不凡的灯饰聚焦聚会气氛，一旁摆上艺术家作品。后方的厨房门片也乘载着数百年的精湛工艺，以手工金箔壁画呈现屋主最爱的花草纹样，欢畅宴饮时，这些令人炫目而精巧的艺术品便成为餐桌上的最佳话题。主卫浴从壁面、洗手台至浴缸，全由具有天然石纹的阿拉伯白大理石打造，天花向上做出圆拱型，相当有欧式宫廷感，并置入一盏由佛罗伦萨采购的编织水晶灯，在洗浴时也能感受到高贵华美的雍容氛围。

COLORS ｜ 色彩

The hostess has a preference for purple and gold, so they are the keynote of the house, representing gorgeous charm. Extending to the back of the living room, there is a three-meter-wide white marble table where the hospitable homeowners entertain guests. Fine artwork can almost be found everywhere in the house. The natural color and simple but magnificent texture makes a luxurious and fashionable New York style house.

　　采用女主人喜爱的紫色和金色，妆点宛如上流贵族般的高雅气韵。由客厅向后延伸，是一张约 3 米的宽敞的白色大理石长桌，让喜欢分享、好客的屋主能尽情款待宾客。家中随处可见值得细细赏玩的高级艺术品。天然的色泽，简约大气的纹理，让豪宅升华进化，跻身为奢华时尚的纽约名流宅邸。

LUXURIOUS STYLE
INTERPRETING HIGH-QUALITY LIFE IN A CONCISE WAY

AN ARTISTIC STATE

艺境

| Location | Kaohsiung, Taiwan
项目地点｜台湾高雄

| Area | 226m²
项目面积｜226m²

| Designer | Jesse Cho
设 计 师｜卓子程

| Design company | Jesse Allen Interior Design
设计公司｜杰西艾伦室内装修设计股份公司

| Photographer | Xinye Liu
摄 影 师｜刘欣业

Main materials | solid wood veneer, stone, leather, glass mirror, cultured stone, etc.
主要材料｜实木皮、石材、皮革、玻璃镜面、文化石等

DESIGN CONCEPT | 设计理念

This residence is located near the harbor without shelters. With good views and sufficient lights, it is a good place for the owner and friends to party. It uses the original low beams as the extension of the ceiling. The margin line makes the vision and kinetonema unity with the seascape. Retaining the original long space and large areas windowing, the designer adopts an open layout to create a continuous and penetrable effect, overlaps and integrates the functions of the space and redefines another lifestyle.

本案坐落于港边，基地条件四周无遮蔽物且视野极佳、阳光充足，是屋主度假及朋友的聚会场所，在设计规划上利用建筑本体的低梁顺应着天花板作线条的延伸，海天一线的边际线手法让视觉和动线与海港景色连为一体；保留原有长型空间与大面积连续开窗的绝佳条件，转换在设计上衍伸出开放式设计手法，创造出连续性与穿透性的效果，将空间机能做重迭与整合，重新定义另一种生活型态。

SPACE PLANNING | 空间规划

The hallway is composed of cultured stones with strong textures, deliberately putting off the space depth and keeping the visual perception briefly. The wall structure is combined with the bookcase in the study. The style of wall-cabinet not only separates hallway from interior spaces, but also increases fun and possibility to the space by getting balance through privacy and permeability. At the same time the iron shelves are embedded in the wall to display the owners' art work and present the owners' taste.

一进门口玄关处是由肌理感极强的文化石所组成的量体，量体感将空间尺度刻意推延，让视觉感受作暂时性停留，墙体结构设计与书房书柜做结合，亦是墙又是柜体的方式，不仅界定了玄关与室内空间，并在隐私和通透间取得平衡来增加空间趣味性与可能性，同时在砖墙上搭配嵌墙铁件层架，让屋主可摆设收藏的展示品，来带出屋主的品位。

The TV wall in the living room adopts small measurement technique, not only broadening the vision, but also increasing the receiving and displaying functions of the other space. The tall cabinet employs solid black iron board and warm solid wood veneers in irregular mode to show the modern style. The landscape area is equipped with couch so as to offer a relaxing space for the owners after work and to view the seascapes outside the window.

客厅电视主墙尺度采用较小量体的设计手法处理，不仅在视觉上产生宽阔感，一体两用的设计也让另一面的观景区空间增加其收纳与展示的机能，旁设的高柜运用实心黑铁板与温润的实木皮柜做结合，以不规则的错落方式呈现交织出现代的摩登风范；观景区以卧榻元素组成起居空间，建构生活的角落空间，让屋主在工作之余可在此作片刻休憩，也让心灵与窗外的海景产生连结。

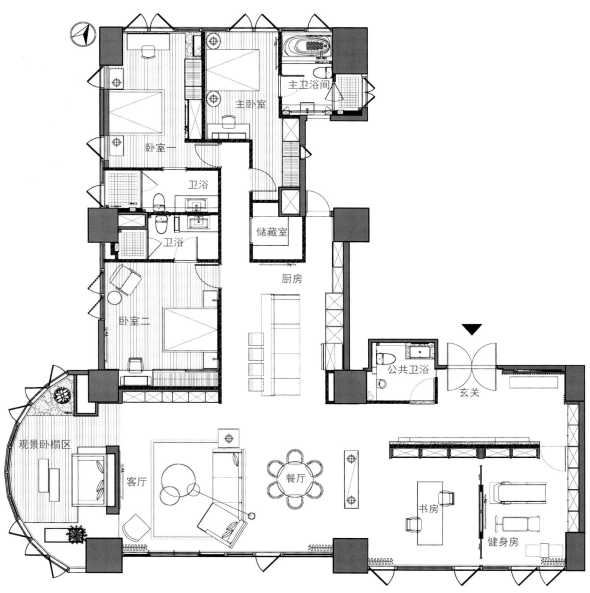

Through the open layout, the beam columns separate the snack bar from the dining table, creating a kinetonema of the principal axis. Under the capacious public space, the structure of the building divides different spaces, such as dining room, living room and bedroom, to increase the elasticity and flexibility of the space. Increasing the proportion of the public areas enlarges the gathering spaces for the family and enhances their relations.

透过开放式的格局，利用梁柱的错落将轻食区吧台和餐桌切割出另一个主轴的动线安排，让敞亮的公共空间在无隔间设计下，透过建筑体的结构错落，区划出餐客厅、卧房区域等不同生活场域，让空间拥有多元使用的弹性与余裕，将公共空间的使用坪数比例增大，全家人团聚的空间可以更加宽敞舒适，亦可以增进彼此感情。

LUXURIOUS STYLE
INTERPRETING HIGH-QUALITY LIFE IN A CONCISE WAY

MOVEMENT AND CONCERTO
乐章·协奏曲

Location | Taipei, Taiwan
项目地点 | 台湾台北

Area | 220m²
项目面积 | 220m²

Designers | Yu Pin-Chi, Tsai Yao-Mou
设 计 师 | 游滨绮、蔡曜牟

Design company | LUOVA Design Co., Ltd.
设计公司 | 创研俬集设计有限公司

Photographer | Jiahe Guo
摄 影 师 | 郭家和

Main materials | wood veneer, ICI, natural marble, titanize, wood floor, etc.
主要材料 | 木皮、ICI、天然大理石、镀钛、木地板等

DESIGN CONCEPT | 设计理念

A movement is composed with melodies. The entire space consists of different parts. An old piano, several old photos, and an Old English style in the memory create a chapter of space. Pure memory abandoned the trifles, bringing out beauty with rationality and sensibility.

"乐章"由数部完整构件的曲目交织而成。一架老钢琴、数张老照片、回忆中的老英式，勾勒出空间的乐谱。纯粹记忆，摒弃琐碎，"美"的原理就是"理性"的展现，与"感性"的理想。

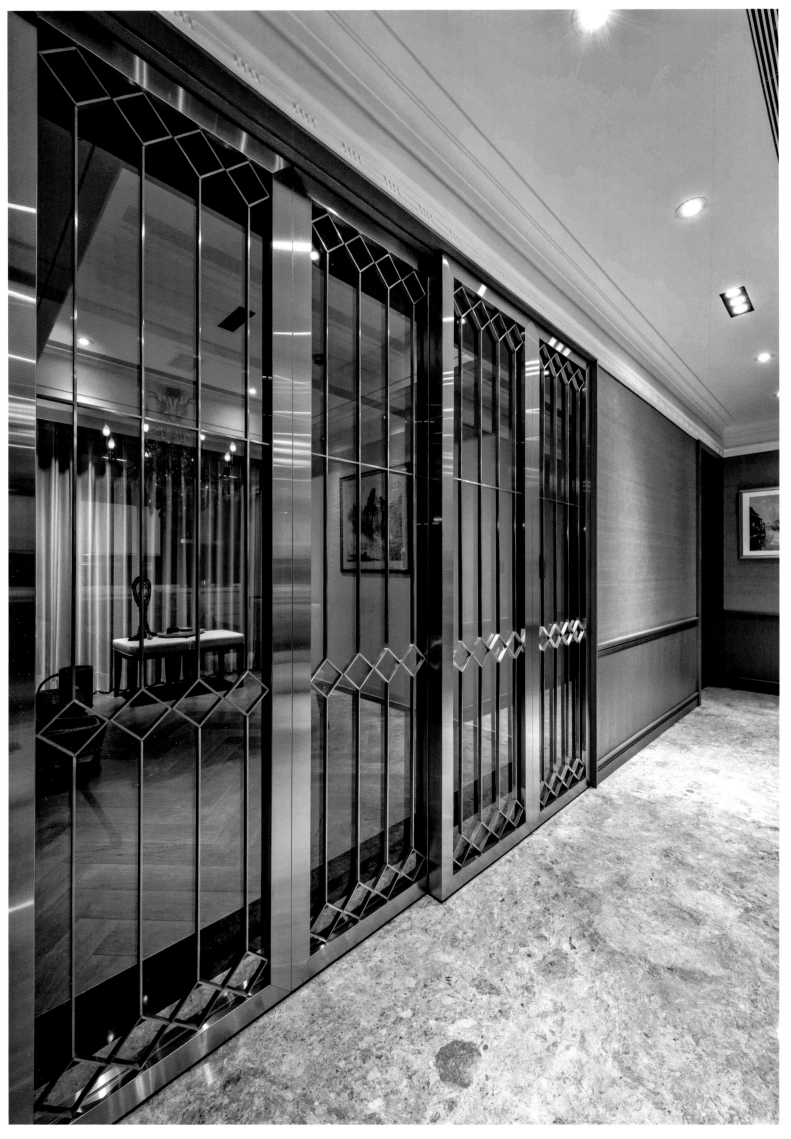

Concerto is originated in the Romantic period; the orchestra is led by the solo instruments, interacting with each other and full of dramatic tension. When listening, you may feel the beauty of balance, unevenness, rationality and sensibility. The melodic space uses classic proportions, making the echoing of color and texture possible. The leading role of the space is highlighted, just like a sheet of music with thoughtful arrangement of melodies. Please listen to the music and enjoy your time.

　　协奏曲，是浪漫主义时期，独奏乐器引领乐队间，似对比实则相互交融的手法，有着十足戏剧张力。聆听般的视觉表现，平衡、不均、理性、感性、冲突美。音符线条般的空间铺陈运用了古典线条比例，颜色共鸣，材质对应，衬托空间的主角，如同一部乐章：起、承、转、合交织着空间。

　　留意，聆听着。

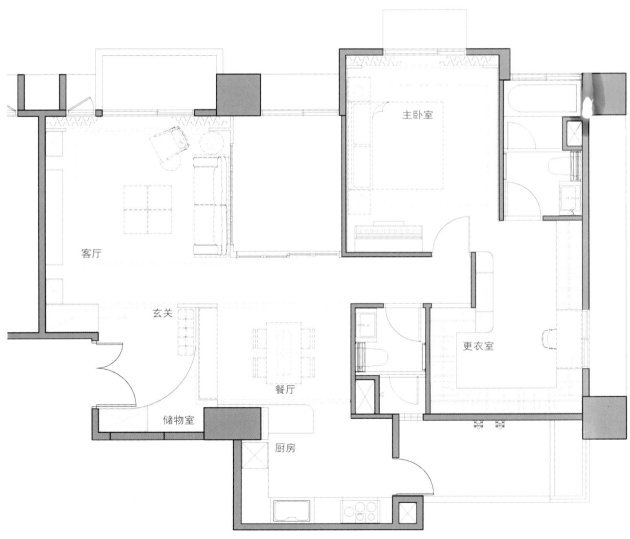

SPACE PLANNING | 空间规划

The arc-shape entrance not only acts as a buffer for vision extension but also provides storage function, making a magnificent and ordered space. The whole space is clearly designed. Living areas such as living room, dining room, multi-function room and cloakroom are included. The elaborate design has made the most use of the room and provides broad view and it is practical, which shows the best aspect of home.

弧形的入口玄关设置，既缓冲了视觉延伸，又兼具大量隐形的收纳功能，空间感大气整齐。整体布局轴线清晰，流畅完善。从客厅、餐厅到多功能室或者卧室衣帽间等生活场所，一应俱全。经过设计师的巧思，横竖线条的勾勒，不仅空间得到充分运用，开阔宽敞，而且实用美观，还原了家最美好的一面。

COLORS | 色彩

The elegance of British luxury is revealed vividly. The gray marble floor, brown leather sofa and green wall processing represent plain British style. Two bright flower paintings are revealing their vigor and vitality on the gray wallpaper.

古典优雅的英式奢华在这里得到淋漓尽致的展现。灰白的大理石地面，棕色的皮革沙发，湖绿色的墙面处理等，每一样都散发着古朴老英式的调性。而色彩艳丽的两幅花朵油彩挂画，在简约灰色的墙壁映衬下，更显生机和活力。

LUXURIOUS STYLE
INTERPRETING HIGH-QUALITY LIFE IN A CONCISE WAY

A RESORT-STYLE RESIDENCE

度假式住宅

Location ǀ New Taipei City, Taiwan 项目地点ǀ台湾新北	Area ǀ 150m² 项目面积ǀ150m²
Designer ǀ Bing 设 计 师ǀ詹秉縈	Design company ǀ S.Z.D Global Design 设计公司ǀ舍子美学设计
Main materials ǀ wood veneer, wood floor with sea-island type, marble, glass, imported wallpaper, etc. 主要材料ǀ木皮、海岛型木地板、大理石、玻璃、进口壁纸等	

DESIGN CONCEPT ǀ 设计理念

As the owner always lives elsewhere, he hopes to stay in a home with hotel-style relaxed atmosphere. The design team chooses stable colors and makes a broad space by concise extension of facade. Each bedroom is designed with hotel-style concept. For example, the thoughtful space for luggage makes you more convenient when packing up belongings. There is a big sitting room in the bedroom, which makes the boundary for the interior and exterior rest areas more obvious. The sitting room is decorated with dark and steady color of low-key texture, but also with hotel-style comfortable atmosphere.

业主长期居住在异地，希望居所拥有像饭店般的质感与放松的氛围，所以设计师在材质上挑选较沉稳的色系，透过简洁的立面延伸，让空间更有宽阔的空间感。在每个卧房都是以饭店房间的概念去做规划设计，例如还会有贴心摆放行李箱的位置，让每次的归来，在整理行李时，不需一直弯腰也不会影响到走道空间。主卧房内含一大起居室，让室内外的休憩区，更有清楚的区分，颜色上也是采用深色稳重的色系去做搭配，带有低调的质感，但也有饭店式的舒适氛围。

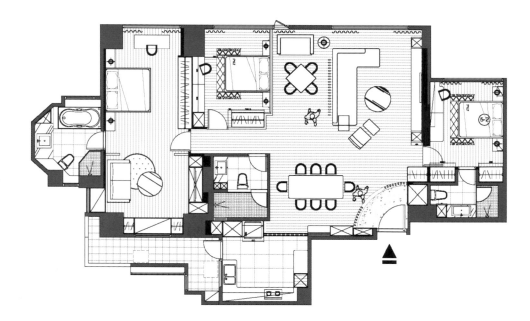

SPACE PLANNING ｜ 空间规划

The ceiling is decorated with recessed lightings; the concise facade becomes an excellent extension; the TV wall is paved with leather marble, making a low-key and comfortable atmosphere. A row of colorful sofas displayed under the cylinder work perfectly in the space, building a rest area. On the dark brown background wall, the visual works are inspired by folding fan. Exquisite aesthetic texture is introduced to the metallic luster on the glossy side, which breaks depressing impression, echoes with the modern lightings and brings different feeling.

天花以两道内嵌照明作为装饰，简洁的立面成为绝佳的延伸，电视墙以皮革面的大理石质材铺陈，使之相融在低调、舒适的整体氛围里；柱体结构的下方陈列一排具饱满色彩的沙发，替空间带来画龙点睛之效，更围塑出后方的休憩空间，而深咖啡色系的背墙上，以扇子开展为灵感的视觉创作，借由亮面的金属光泽，引入细腻的美学质感，打破沉闷的印象，且与现代感的设计灯饰相互呼应，烘托一抹新意。

COLORS | 色彩

The interior design is decorated with international matching skills; low-key and calm color can be found in the space. The designers first lay the keynote of the whole style, and then choose the selected furniture, soft decoration and aesthetic lines, making the artistic vision and color become the highlights. Stepping into the entrance, the curved lines and exotic material set the interior and exterior space apart. Following the round stone floor, passing through the grille, you will be attracted by the bright purple in the living room area.

室内选择国际的搭配技法，低调、沉稳的色系游走在空间里，先为整体的室内风格定调，透过精选的家具、软饰与美学线条装点其中，让艺术感的视觉与色彩作为绝佳的亮点。来到玄关，透过弧形的线条与异类质材自然区隔内外领域，而沿着圆形石材的地面，透过格栅，视线被客厅区域亮紫色系的沙发而吸引。

THE ELEGANT RESIDENCE IN A HOTEL STYLE WITH EMOTIONAL LIGHTS

日光叙意 饭店式的品位居宅

Project name ∣ Xie Residence in Hsinchu City 项目名称∣新竹谢宅	Location ∣ Hsinchu City, Taiwan 项目地点∣台湾新竹	Area ∣ 248m² 项目面积∣248m²
Designers ∣ Junsong Yang, Youcheng Luo 设 计 师∣杨竣淞、罗尤呈	Design company ∣ Ahead Design 设计公司∣开物设计	Photographer ∣ Sam Cen 摄 影 师∣岑修贤
Main materials ∣ stone husk, tawny glass, wood veneer, wallpaper, stone paint, stone, etc. 主要材料∣石皮、茶镜、木皮、壁纸、石头漆、石材等		

DESIGN CONCEPT ｜ 设计理念

Interpreting the aesthetics of life and presenting the appearance of the space in a new era, the designers create a fresh and elegant residence by modern hotel-style exalted textures, lightly sparking lights, strictly chosen materials and lines and surfaces layout.

演绎生活美学，铺述新时代的空间样貌，设计师以现代质韵糅合饭店式的尊贵质感，透过光线轻洒、严选材质及线面铺陈，将场域形塑成清新优雅的品味宅邸。

This project is a new home with a clear layout. The designers scrupulously master the structure and proportion of the space. They retain the original layout and define the kinetonema by slight change to promote the relation between resident and environment to achieve coordination and dialogue. Slowly walking in, the private space is capacious and bright; the living room and dining room are maximized to connect with each other. By using neat lines, stones and tawny glasses, it looks more beautiful. The recombination and diversity make people relaxed and carefree.

本案为脉络清晰的新成屋，设计师严谨掌握空间结构与比例，保留原始格局的优势，仅透过部分微调来定义动线的关系，让居住的人与环境达到协调的关系与对话。缓步入内，宽敞明亮的公共场域轩宇大气，客、餐厅以最大化来串联共生，更借着利落线条及石材、茶镜的搭配运用，呈现视觉上的交织美感，其复合性与多元样貌，更让返家的人们得到悠然舒心的享受。

DECORATIVE MATERIALS | 装饰材料

The designers create large scales of public area, combine consummate execution with aesthetic designs and blend in with a hotel style. The pictures next to the window follow the owner's mood and preference. Wherever to enjoy the pictures, you can feel the charm of art. The rough wood veneers are used in the TV wall to create a grand appearance. The cabinet next to the entrance is linearly cut to make the space neat and clean. The surface of the cabinet chooses clipping stones to foil the wonderful area. The walls of the dining room and entrance are covered with stone paint to create a visual effect just like raw stones. The walls of the master bedroom use colored wood to improve the cozy atmosphere in the space. Above the bed are chocolate bricks, which contain contemporary and creative design elements.

营造大尺度的公共场域，以精湛的施作功力与设计美感，融入饭店式的大气风范。窗边的挂画门片依照屋主的心境与喜好，恣意移动观看的欣赏位置，用艺术挥洒千万的风情。运用粗犷肌理的石皮，作为电视主墙的立面表现，围塑恢弘大气的面貌。入口处旁的储物柜以线条式切割，让整体利落且干净，中间的柜面则选用拼接的石材来映照场域的丰富精彩。餐厅主墙与入门处立面使用石头漆来铺述，营造达到与原石相仿的视觉效果。主卧墙面运用染色的绘木，挹注卧眠空间舒适的温度；床头则以巧克力砖的造形，形塑出现代感与创意兼具的设计元素。

SPACE PLANNING | 空间规划

Based on the same design axis, the sleeping area makes full use of the spatial aesthetics to present interweaving aesthetics. The designers use modern fashionable words to endow every space with a different appearance to create the best sleeping atmosphere. The sliding doors with straight grains divide the functions of the bedroom and dressing room. Its texture with penetrable capability brings lights into rooms. Simple, young and fashionable designs describe the girl's personality and spirit. The kid's companion space also has basic functions as public area, with creative lines and modern lamp decorations and software. The design of dining room has more free space as an elastic space; you can choose everywhere to sit, chat or read.

卧眠的私人场域，则环绕相同的设计主轴，充分利用空间美学去体现交织的美感，并以现代时尚的语汇赋予每间房间不同面貌，勾勒出最优质的寝眠环境。直纹玻璃拉门划分了卧房与更衣室的机能关系，其带有穿透感的质材特性，也将光线引入其内。简约而带有年轻感的时尚设计，轻轻谱出属于女孩的个性与神采。属于小朋友的交谊空间，同样拥有公共场域的基本机能，更多了创意线条与现代感的灯饰与软件。餐厅在设计上，给予相当自由的空间设计，同时也作为一处弹性空间，可以自由选择任何角落，随性的坐着、谈天或阅读。

245

室内绿化
INTERIOR VIRESCENCE

自然采光
NATURAL LIGHTING

装饰材料
DECORATIVE MATERIALS

设计理念
DESIGN CONCEPT

色彩
COLORS

引景入室
BRINGING SCENERY INTO HOUSE

空间规划
SPACE PLANNING

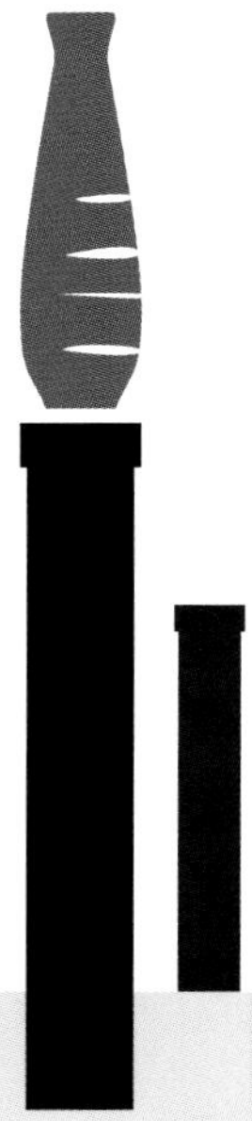

工业风

以现代前卫的材质和布局打造loft零压居住空间

INDUSTRIAL STYLE

CREATING A PRESSURELESS LIVING
LOFT BY MODERN AND FASHION
TEXTURE AND LAYOUT

INDUSTRIAL STYLE
CREATING A PRESSURELESS LIVING LOFT BY MODERN AND FASHION TEXTURE AND LAYOUT

GATHERING

聚

Location ｜ Yilan, Taiwan 项目地点 ｜ 台湾宜兰	**Area** ｜ 436m² 项目面积 ｜ 436m²	
Designers ｜ ChiChi Chiang, Hank Hsiao 设 计 师 ｜ 蒋孝琪、萧明宗	**Design company** ｜ Yi Ciao Interior Design 设计公司 ｜ 逸乔室内设计	**Photographer** ｜ Yvonne Kao 摄 影 师 ｜ 高伊芬

Main materials ｜ iron rust paint, cultured stone, natural wood veneer, natural rock slices, oak floor, etc.
主要材料 ｜ 锈铁漆、文化石、天然木皮、天然岩片、橡木地板等

DESIGN CONCEPT ｜ 设计理念

It is in the motivation of having a place to live in the old age that the homeowner bought this villa in Yilan. He hopes to have such a place to live and get together with family members during festivals as well as entertain friends. In this four-story house, different functions and designs are applied in each story to satisfy the demand of each family member.

老有所归，这是屋主当初买下这栋位于宜兰透天别墅的动机。屋主希望退休后自住及家人返乡过节时亦能作为招待众多好友们的场所。在这四层楼的空间里，为了满足家族成员的喜好与习惯，在每层楼皆规划了不同用途与细腻的设计。

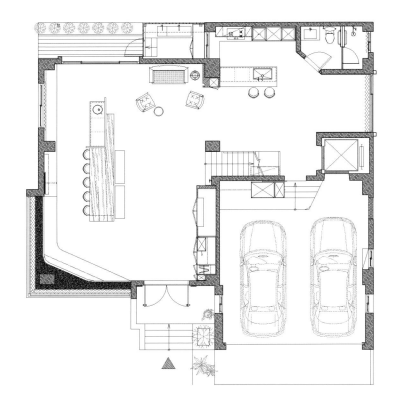

The whole house is designed in an industrial style with simplicity and openness. The white or gray concrete wall brings a nostalgic but modern visual effect, at the same time creating a modern style with simplicity and calmness. With innovations, ingenuity and simple texture, the designers create a random life fun.

整体风格以简约开放式工业风设计为主线，不管是涂上白色或是灰色的水泥墙壁，都能带给室内一种怀旧却又摩登的视觉效果，同时创造一种素朴的沉静与现代感。通过设计师的创意，对生活的巧思、以及材质本身素朴的质感、打造出一种信手拈来的生活趣味。

SPACE PLANNING ｜ 空间规划

The living room and dining room are planned in an open way, where the light colored wood floor matches perfectly with the simple wall. In order to make a space with rich layering, the designers use different materials in different parts. For example, the grained brick wall on the back of the sofa and the yellow painted glass add some uniqueness in the space. The dinner table is skillfully connected to the island, making a broader space for dining and gathering to satisfy the owner's demand to entertain friends.

客厅与餐厨领域，整区采取开放式规划，浅系木地板与给人简约感的清水模墙、柱相辅相成。为了使空间更具层次变化，设计师特别在部分区块注入不同的素材点缀，像是沙发后方整面的木纹砖墙与厨房区的黄色烤玻，皆为该区增添不少特色。同时餐桌巧妙地连结中岛，将用餐与聚会空间再次放大，以满足屋主招待朋友之用。

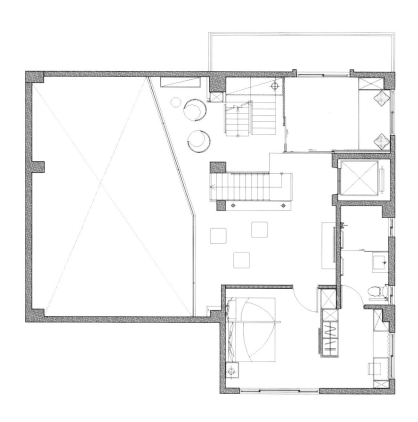

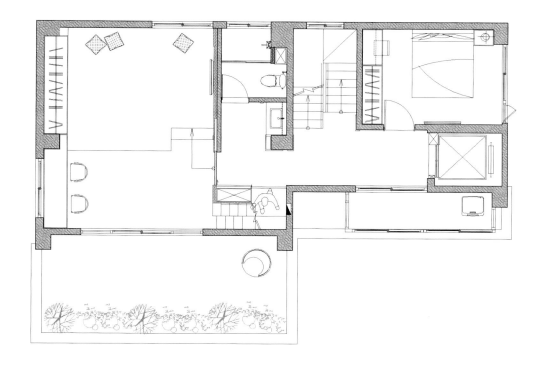

NATURAL LIGHTING | 自然采光

In the exterior of the master bedroom, there is a big terrace. After replanning, some parts of the terrace are transformed into a sitting room. The unique scuttle provides wonderful lighting. It enables the owner to enjoy the sunshine whether in the balcony or in the sitting room.

　　主卧外部原为大型的露天阳台，经重新规划后将部分空间调整成起居室，加上独特的天窗设计，给室内提供良好的采光，让屋主不论是在外阳台或是内起居空间皆能享受到日光沐浴的美好。

INDUSTRIAL STYLE
CREATING A PRESSURELESS LIVING LOFT BY MODERN AND FASHION TEXTURE AND LAYOUT

SOJOURN
旅居

Location | Taipei, Taiwan
项目地点 | 台湾台北

Area | 159m²
项目面积 | 159m²

Designers | Shin-Jie Lin, Ting-Liang Chen
设 计 师 | 林仕杰、陈婷亮

Design company | Ganna Design
设计公司 | 甘纳设计

Photographer | MWphotoinc / Siew Shien Sam
摄 影 师 | MWphotoinc / Siew Shien Sam

Main materials | wood veneer, spray paint, glass, iron part, wear-resisting wood floor, ceramic tile, etc.
主要材料 | 木皮、喷漆、玻璃、铁件、超耐磨木地板、瓷砖等

DESIGN CONCEPT | 设计理念

After having a good understanding of the home owner, the designers create an ideal residence based on the life blueprint. Located on the tranquil mountainside of the suburb Taipei, this residence boasts clean air and comfortable outdoor surroundings. Simple but fashionable black, white and gray colors are used in its interior design matching bright colored furniture, which enables the owner to enjoy himself in leisure times.

设计师在充分了解屋主入住需求后，依据生活蓝图来构思理想的栖身住所。住宅选择落脚台北近郊倚靠山腰的清幽环境，空气清新，户外环境自然恬适。室内选择简约时尚的黑白灰色调，佐以颇有设计感的亮色家具增加空间色感，让居者工作之余可以身心放松在此享受闲暇居家时光。

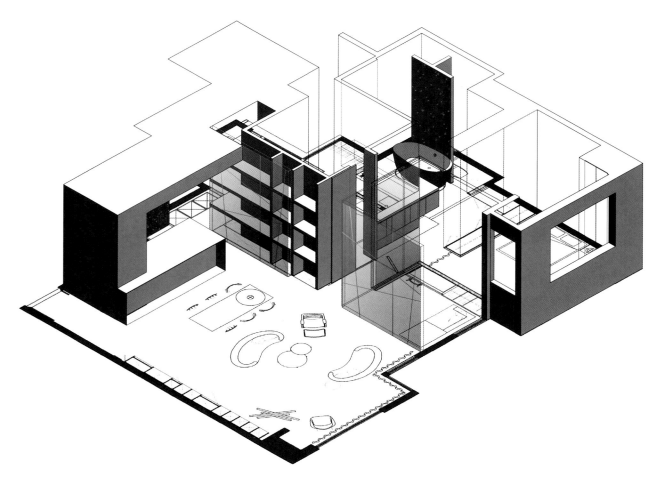

SPACE PLANNING | 空间规划

Taking the owner's habit into consideration, the boundary of the living room and dining room is in an open style with open horizons and convenient movement. The space planning in the master bedroom is just like the arrangement of a hotel to interpret the relation of each area and divide their scopes, which provides an unfettered life.

依照屋主日常习惯，客厅与餐厅分区为开敞式，视野开阔，行为动线流畅无阻。主卧空间采以饭店式的格局安排，来诠释区块使用的行为关系，渐进划分各机能范围，经由回字状动线，自在享受无拘的休憩生活。

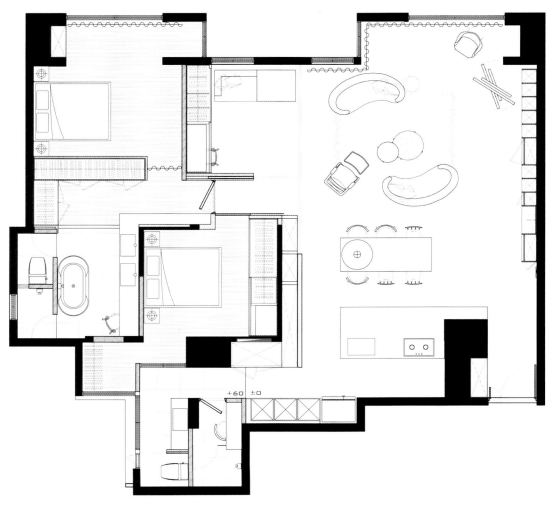

COLORS ｜ 色彩

The main colors of the space are black, white and gray. The island shaped kitchen takes black color as backgrond, and matches the wood storage wall. A big slide wall made of black iron part can be moved to the working area and functions as a partition to serve for the guest room. The dark blue display wall shows its unique function after releasing from the rigid pattern of TV. It can change the cells gradually with regular order, creating an interesting storage idea in the changes, adding some colors in the room while taking the storage demand of entrance and books into consideration.

以黑白灰色为整体主调，中岛型厨房取黑色为背景，搭配一旁木制收纳墙，透过一道黑色铁件所订制的大滑门，可移动至工作区为隔墙另作客房之用。而深蓝色展示主墙，在减去电视机的拘谨模式后，让不规则大小格状延展渐变，不失规律的秩序性，营造趣味转换下的收纳巧思，并整合玄关、书籍应有的置物需求，增添空间单纯少有的色彩亮点。

INDUSTRIAL STYLE — CREATING A PRESSURELESS LIVING LOFT BY MODERN AND FASHION TEXTURE AND LAYOUT

THE WALL

一道墙

Project name ｜ Ju He Fa Tian Building Showhouse 项目名称｜聚合发天厦实品屋	Location ｜ Taichung, Taiwan 项目地点｜台湾台中	Area ｜ 149m² 项目面积｜149m²
Designer ｜ Shuangqing Lin 设 计 师｜林双庆	Design company ｜ KEI Design Studio 设计公司｜界境设计	Photographer ｜ Junjie Liu 摄 影 师｜刘俊杰
Main materials ｜ carved white marble, wood veneer, gray mirror, titanize board, leather, iron part, etc. 主要材料｜小雕刻白大理石、木皮、灰镜、镀钛板、皮革、铁件等		

DESIGN CONCEPT ｜ 设计理念

Building a spatial wall between the low-key and tranquil pure white can provide a poetic layout.
Building a visual wall between the rhythm of mixed real and virtual can provide a picturesque mobility.
Building a living wall between the dialogue of straight line and oblique line can provide a timeless performance.
The original four rooms integrate the study into the public space, bringing light and beautiful views into the room. The axis and wall arranged the space of entrance, living room, dining room and study. The mobile axis integrates life and light, bringing a comfortable living space.

　　立一道空间的墙，在低调静谧的净白之间，层次建构如雕塑般写意。
　　立一道视觉的墙，在虚实交织的律动之间，内外流转如风景般瑰丽。
　　立一道生活的墙，在直线斜线的对话之间，交迭展演如画作般隽永。
　　原本四房的空间，将书房以开放姿态融入公共空间，将光线及高楼层的窗景纳入室内。空间的机能以轴线与造型墙将玄关、客厅、餐厅、书房界定围塑。动线轴串连了空间生活，纳入光影，生活的温度在其中蕴酿扩散。

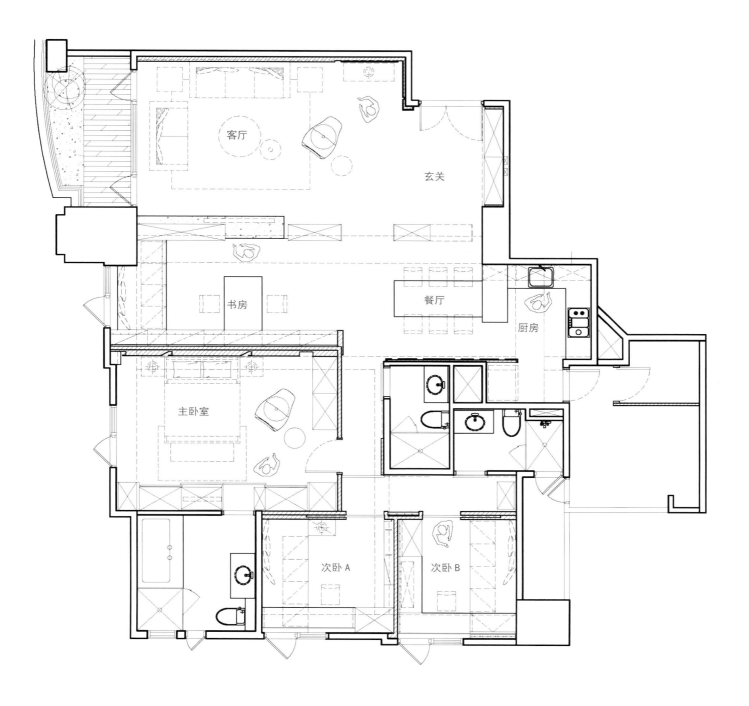

NATURAL LIGHTING | 自然采光

Massive panoramic French windows offer a comfortable experience when the sunshine penetrates the white curtain. Considering its high level, the house enjoys a broad vision with beautiful scenery outside the window. In the meantime, the irradiated light is reflected by the marble wall, mixing real and virtual, hiding scenery in the scenery.

　　大面积全景的落地窗，虑过纯白色的纱幔，阳光透射进来，温暖安静。由于楼房本身的层高，窗外景致迷人，视野开阔。而照射进来的光线，经过大理石墙面的反射，虚实交错，景中藏景。

SPACE PLANNING | 空间规划

Under the linear cutting of the diagonal lines, the wall presents its shape and texture, defining the spatial properties ingeniously and at the same time, maintain the possibility of spatial dialogue. The extension of lines brings mobile space and the tensility of space follows the pattern of people, life and light, making life full of possibilities.

造型墙在空间对角线的线性切割下，以虚实的量体、材质雕塑呈现，巧妙地定义空间属性又保留空间对话的可能性。量体的斜向延伸衍生出穿透而律动的空间感，空间的张力依循着人、生活、光影的交迭，让生活有了无限可能。

INDUSTRIAL STYLE
CREATING A PRESSURELESS LIVING LOFT BY MODERN AND FASHION TEXTURE AND LAYOUT

THE BEAUTY OF MINIMALISM

极简之美

Location ｜ Kaohsiung, Taiwan 项目地点 ｜ 台湾高雄	**Area** ｜ 258m² 项目面积 ｜ 258m²
Designer ｜ Yuwei Lin 设 计 师 ｜ 林宇崴	**Design company** ｜ Platino Interior Design 设计公司 ｜ 白金里居空间设计

Main materials ｜ marble, rotating shield, strengthened tawny glass, artificial turf, etc.
主要材料 ｜ 大理石、旋转隔屏、强化茶玻、人工草皮等

DESIGN CONCEPT ｜ 设计理念

In the city with beautiful sunshine, the light and space make a series dialogue which brings a modern bedroom with minimalist style. The design team takes the owner's aspiration for broad vision and relaxed atmosphere into consideration. They bring living room, study, kitchen, dining room and multi-function room together to make a broad public area. The white rotating shield, the island and the sliding-glass door divide each area to maintain the function of each room and the potential extension of the demand.

阳光总是美好的港都城市里，光线和空间谱成一系列的光影对话，温暖了现代风的极简居室。我们聆听完屋主对于开阔的向往，以及工作之余需要极度放松的精神诉求。将客厅、书房、厨房、餐厅和多功能室都纳入公共领域，让公共空间更为开阔。再以白色旋转隔屏、中岛、玻璃拉门划分场域，保有该区域的功能性和未来人口增加的扩充性。

283

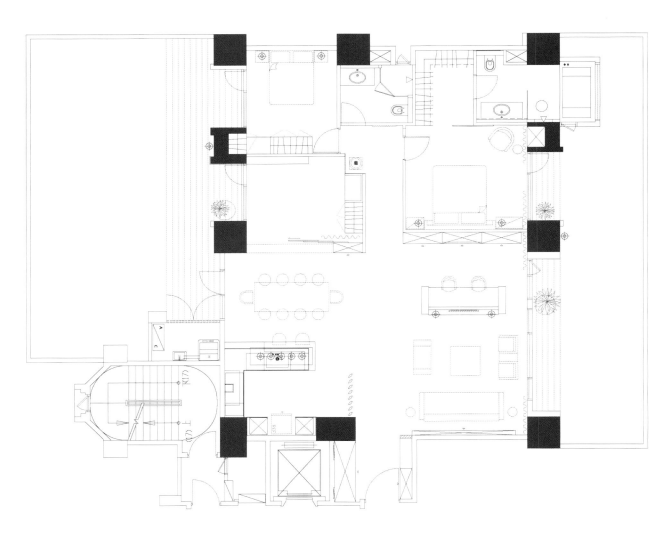

SPACE PLANNING | 空间规划

In the large space, there is a big area of terrace to relax yourself. No complicated elements are found in its interior design; the designers have a preference for simplicity. Artificial turf can be found on the floor of the living room, creating an urban oasis when matched with the mild plastic synthetic wood floor. It is a good place for walking, having a party, gardening and enjoying leisure time. The dining room with excellent daylight enables people to enjoy the green scenery outside the window and have a good mood while having meals.

偌大的空间里，有一大部分是可以舒展身心的露台。室内没有繁复的元素堆砌，设计师尊崇干净利落的简约感，客厅地面铺上仿真度颇高的人造草皮，搭配朴实温润的塑合木地板，仿若城市绿洲。在这里可以散步运动、举办户外派对、父母种花莳草、修剪败叶、尽享闲暇时光。采光极好的餐厅让居者用餐时，一眼览尽户外绿色风景，开启用餐好心情。

INDUSTRIAL STYLE
CREATING A PRESSURELESS LIVING LOFT BY MODERN AND FASHION TEXTURE AND LAYOUT

TRUE HOME
身心居所

Location ｜ Taichung, Taiwan 项目地点｜台湾台中	Area ｜ 415m² 项目面积｜415m²	
Designer ｜ Yu-Tang Chen 设 计 师｜陈煜棠	Design company ｜ Unison Space Design 设计公司｜共禾筑研设计有限公司	Photographer ｜ Kevin Wu 摄 影 师｜吴启民

Main materials ｜ wood veneer, stone, glass, wood board, titanium plate, leather fabric, imported decorating fabric, etc.
主要材料｜木皮、石材、玻璃、木饰板、镀钛板、皮革布、进口家饰布等

DESIGN CONCEPT ｜ 设计理念

A house is not a home unless it contains temperature and love. In this project, the designer connects the whole family's life and offers exclusive warmth of the family by putting different design elements in the space.

一个没有温度与爱的屋子，称不上是家。在此案例中，设计师透过设计元素在空间中的流转，串起一家人的生活范畴，也串出专属于家该拥有的深刻暖度。

According to different living habits of the family members, this five-story building with three generations needs to be redistributed, especially the functions of every floor. The basement is mainly used as a garage. Besides parking two cars, there is enough space to store a lot of sundries. Thereby the daily activity areas are separated and the living areas are clean, neat and comfortable. The public area in the first floor is an important place to be together. So the living room and dining room use open-mode design in order to give the family a bright and broad activity area. The design of the table in the kitchen combines with the function of a bar. So it is placed aslant to relieve the pressure from dining table and the stairs, which creates fluent working lines. Other floors are private bedrooms for the elders, the young couple and their children. Focusing on briefness and comfort, the rooms provide relaxing atmosphere for the family. Every bedroom has a bathroom. All the family can enjoy a complete and convenient life.

　　三代同堂的五层楼透天老屋，依照居住者的起居习惯，重新分配每层楼的格局机能，明确界定出地下一楼主要作为车库使用，除了停放两台汽车之外，尚有足够的空间可储放大量杂物，借此与日常活动的场域做出区隔，让起居空间保有干净整洁的舒适环境。一楼的公共空间为家人间联系情感的重要场所，因此客、餐厅采用开放式设计，给予一家人明亮宽适的活动场域。厨房的餐桌结合吧台机能，并特别以斜角安排，巧妙使餐桌与楼梯间的压迫感得到缓冲，创造出流畅的行走动线。二到四楼主要则分别为长辈、男女主人与小孩子的私人卧房，以简约舒适为题，营造令人放松的休憩氛围，每间卧房并设有独立卫浴机能，让屋主一家人都能享有完整且便利居家生活。

DECORATIVE MATERIALS AND SPACE PLANNING | 装饰材料、空间规划

The TV wall was made of Scotland marbles, which creates a restrained and decent identity. The changes of ditch and gap lines add a sense of stratification. The dark wood textures decorate in the dining room to warm and color the space. The door in the guests' bathroom uses the same technique to be more beautiful and latent. The design of dining table combines with the function of a bar. It is placed aslant to relieve the pressure from dining table and the stairs. The design of layer combined with mirror adds visual fun to the space and enlarges the space at the same time. The neat lines and transition variation are used to decrease the visual height difference of the ceiling and to become a modern design.

　　以苏格兰灰大理石作为电视主墙，营造内敛大方的立面表情，并借由沟缝线条的变化，增添视觉层次感。而深色木质元素点缀在餐厅段落，为空间添色加温，客用卫浴的门片也采用相同手法将其美化、隐藏。将餐桌结合吧台，并施以斜角设计，化解餐桌与楼梯之间的压迫感。以镜面结合表示层架的设计增添视觉的趣味性，同时放大空间感。利用利落的线条与转折变化，修饰天花板高低差的视觉差距，同时也成为具现代感造型设计。

COLORS | 色彩

The designer uses simple plain wallpaper to decorate the wall behind the sofa and intersperses simple funny artistic work on it, which is elegant and exquisite. The wall in the master bedroom also adopts simple restrained surrounds to create a quiet atmosphere. In particular, the back of the mirror on the right of the bed is the window. The designer uses mirror sliding doors to be more beautiful and to maintain advantage of the original lighting. The elders' rooms focus on comfort. The simple and elegant tones combine with the warm wood, which creates feelings of leisure and relief. The clean and bright children's rooms highlight children's innocence and are partially decorated with earth colors to warm the spaces.

设计师以简约的素色壁纸作为沙发背墙，简单点缀上趣味的艺术画作，形塑雅致品味。同样采用简约内敛的灰围塑主卧室的安定氛围，特别的是，床头右侧的镜面后实为窗户所在，设计师利用镜面拉门的手法将其美化，也同时保留了原有的采光优势。长亲房以舒适为题，素雅的色调结合温润的木色，营造休闲舒压之感。而纯净明亮的小孩房，突显孩子们的天真单纯，局部利用大地色系做点缀，为空间增添温润暖度。

INDUSTRIAL STYLE
CREATING A PRESSURELESS LIVING LOFT BY MODERN AND FASHION TEXTURE AND LAYOUT

FLEETING TIME

光影流年

Project name ｜ Mountain and Stone Assemble 项目名称｜璞石山汇	Location ｜ New Taipei City, Taiwan 项目地点｜台湾新北	Area ｜ 297m² 项目面积｜297m²
Designers ｜ Zhixiang Lee, Zixian Liu, Ruiwen Guo 设 计 师｜李智翔、刘梓娴、郭瑞文	Design company ｜ Waterfrom Design 设计公司｜水相设计	Photographer ｜ Sam Cen 摄 影 师｜岑修贤

Main materials ｜ appeared stone, corrugated wood veneer, walnut solid wood floor, iron part, wood cultured stone, stone floor, wood block, carpet, etc.
主要材料｜斧劈面锈石、波纹木皮、胡桃实木地板、铁件、木文化石、石材地板、木纹砖、编织毯等

DESIGN CONCEPT ｜ 设计理念

"I am forever chasing light. Light turns the ordinary into the magical." — Trent Parke

Light is the essence of photography and also reflects the basis of reality. The basis makes the designers think about the relative space: under the light, the space has its expression and the meaning of time. The owner is an amateurish photographer and is sensitive to light and shadow. How to create a perfect and reasonable home of light and shadow is what he concentrates on especially. The biggest characteristic of this project is management and application of light and shadow. The designers focus on making the influence of light and shadow on the space maximized. Feeling the flowing of light and shadow and regarding the space as a big camera, every second of life is creating every decisive moment.

"我不断追逐着光,光能将平凡的东西化为神奇。"——Trent Parke

光线是摄影的本质,也是反映物体现实的基础。此基础让我们反思与之相对应的空间:在光线下,空间有了表情,也才有了时间的意义。业主是一名业余的摄影师,所以对光影特别敏感,如何创造出一个完美合理的光影之家是业主所特别关注的。本案最大的特色就是对于光线的处理和运用,我们专注让光线在空间的影响力极大化,感受光线在空间流动,视空间为一台巨型相机,生活的每一秒,都在创造每个决定性的瞬间。

NATURAL LIGHTING | 自然采光

The workspace in the basement is lack of natural light. So the designers create six iron part punched cabinets by borrowing things to draw pictures, and make the artificial light shuttling back and forth in the cabinets as if the lights are entering into the aperture, creating an effect different from natural light and shadow.

The second floor is the master bedroom. The designers use wooden grating and iron part to lead the natural light into the center of the space and create ways to lead different lights, such as time span, angle and direction. It is the trickling and warm lights that flow over the grating. The gallery which displays artworks and books connects the master bedroom with the dressing room.

The lights enter not only through the French sash, but also through the wall control and management to change the original way. The entrance of the study on the third floor is designed deliberately to control the amount and angle of the lights, to improve the orthoptic privacy and to keep the natural lights change.

地下室的工作空间因自然光线的不足，所以我们以借物画影的方式创造6座铁件冲孔的柜体；让人造光源穿梭在柜体间犹如进入镜头光圈的通光量，产生不同于自然光影的效果。

二楼空间为主卧室区域，我们用木格栅与铁件将自然光源导入空间的中心，创造不同光线进入的方式，例如时间长短、角度方向，是细流的光，是暖洋的光，在格栅间流转。展示艺术品或书籍的廊道，连接主卧室与更衣空间。

光线不仅仅是大面积的从落地窗进入，而是透过墙体控制并处理，改变印象中观看的方式。三楼的书房空间，入口设计刻意控制自然光的量与角度，改善廊道视觉直视的隐私性外，并让自然光保持变化。

BRINGING SCENERY INTO HOUSE
引景入室

"When you are shooting, try to add a box in front of the scenery, you will make the ordinary scenery extraordinary." — Ansel Adams

Using the photographic framing theory, the long public space of the gallery on the first floor passing through two cabinets in the living room forms a frame. The window behind the cabinets leads in natural lights and the green scenery. The place of the frame is the scenery.

"当你拍摄风景照时，试着景物前加一个框，你会让平凡的景物变得极不平凡。"——安瑟·亚当斯

借用摄影的框景理论，一楼长型公共空间的廊道透过客厅两座柜体形成一道框景，柜体后方开窗导入自然光源与梯间的大开窗端景绿意，框处即景。

INDUSTRIAL STYLE — CREATING A PRESSURELESS LIVING LOFT BY MODERN AND FASHION TEXTURE AND LAYOUT

THE WHITE RESIDENCE

白色大宅

Location ｜ Zhubei, Taiwan
项目地点｜台湾竹北

Area ｜ 283m²
项目面积｜283m²

Designer ｜ Hsu Wei Yi, Hguan design team
设计师｜徐玮逸、禾观设计团队

Design company ｜ Hguan Interior Design
设计公司｜禾观空间设计

Main materials ｜ marble, stainless steel, weathering slate, ore block, white paint, kieselguhr, wear-resisting wood fllor, etc.
主要材料｜大理石、不锈钢、风化板、矿石板、白色烤漆、硅藻土、超耐磨地板等

DESIGN CONCEPT ｜ 设计理念

This project is a new home for a young couple. The designers form an open setup for the space, blend in many storage functions, present cultural elements to beautify the space and create a beautiful, practical and bright residence.

Entering the door, the white cabinet displays art works in the hallway. The stainless steel arch door makes a first impression of cultivation. The hall is based on a white tone, covered with earth color patterned marble floors, and uses noble materials to lift the textures. The elevated recreational area adopts open and warm wood textures to create a relaxing atmosphere. The hidden lift table under the floor provides practical functions. The walls are decorated with display cabinets made by wood boards and white baking steel to add rich textures.

　　此案是一对年轻夫妇的新居场域，设计师替空间型塑开阔格局，同时融入大量收纳机能，并展示艺术文化元素美化居家，打造出一间兼具美观与实用的敞亮大宅。

　　走进大门，玄关规划雪白柜面妆点展示艺品，采以不锈钢塑造门拱线条，表达充满艺术文化涵养的第一印象。厅区整体则以白色调做基底，并铺述大地色花纹大理石地面，以高贵的建材质地提升空间质感，一旁的架高休闲区，则以开放形式与温润木质建构放松感受，并于地板底部暗藏升降桌，提供生活中的机能使用。立面则以木作层板穿插白色钢烤做出展示柜，替墙面增添丰富表情。

DECORATIVE MATERIALS | 装饰材料

Abandoning the original polished tiles, the public areas adopt marble textures to create a low-key atmosphere. As for the decoration, it uses Nordic elements, such as solid wood veneers and weathering slates, to warm the space. Besides, the good lighting turns the low-key space into a warm, grand and cozy living space. The space becomes a comfortable and creative residence rather than a single style space.

  公共区域退除建商原有抛光地砖，采用更具质感、大气的大理石石材地面，营造出低调奢华的意象。但在装潢设计上却采用大量实木皮、风化板等北欧风格元素让空间附带了一些暖意，加上本身屋子采光良好，使这原本低奢的空间摇身一变转化成了阳光温暖、质感大气、温馨休闲兼顾的居家空间。空间不再是以单一风格为主轴，而是舒适且充满创造性的住宅空间。

SPACE PLANNING | 空间规划

Functionally, the double flipping television design of the TV wall in the living room increases the flexibility of the space. No matter you are in the living room or the tatami room, you can watch the TV easily. The deep sofa provides a good place for the owners to relax and promotes the leisure atmosphere of the space. The dining room is divided into snack area and cooking area without functional conflicts, improving the utilization of the space. The big dining table in the center of the dining room highlights the grand atmosphere of the space. The dining room has a broad area and a black mineral display cabinet. The black setting foils the elegant posture of the exhibits. At the same time the wall is decorated with clear wood texture, which manifests the visual extension.

机能上，客厅电视墙双面翻转的电视设计，不论在客厅或和室都能轻松观赏电视，让空间使用上更自由灵活。而座深较深的沙发则是提供屋主放松休息的好地方，同时也提升整体空间休闲的氛围感。厨房区则采用轻食区与热炒区分隔，在使用上互不冲突，空间利用更完善。而餐厅区正中央的一张大餐桌，则突显出这个空间大气感。餐厅则给予开阔的宽广场域，并规划黑色矿石板展示柜，透过黑色背景衬托展示品的优雅姿态，同时以清晰的木作纹理铺陈于墙，彰显视觉上的延伸层次！

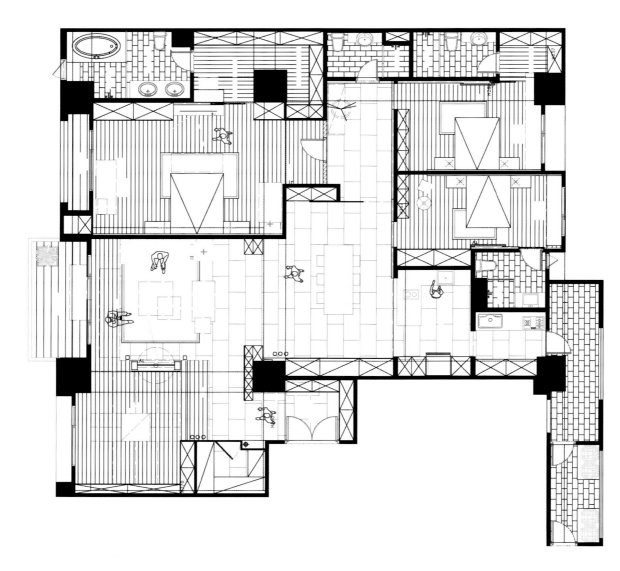

COLORS | 色彩

The bedroom is in simple hoary color with comfortable textures. In addition, the display platform of the bedroom background shows the beauty of modern and concise lines. At the same time, the arrangement and color of the chair and other decorations make the space and software complement each other, performing a clean, neat and simple picture.

卧房延续整体案例色调，简约的灰白色，质感舒心。此外于床头背景加入展示平台，展现简约的线条美，同时透过单椅、家饰等合宜布置与色调安排，让软饰与空间相得益彰，勾勒出干净利落的极简画面！

图书在版编目（ＣＩＰ）数据

台风台韵：解读台式新空间美学 / 深圳视界文化传播有限公司编．-- 北京：中国林业出版社，2016.5
ISBN 978-7-5038-8541-9

Ⅰ．①台… Ⅱ．①深… Ⅲ．①住宅－室内装饰设计－中国－图集 Ⅳ．① TU241-64

中国版本图书馆 CIP 数据核字（2016）第 103226 号

编委会成员名单
策划制作：深圳视界文化传播有限公司（www.dvip-sz.com）
总 策 划：万绍东
责任编辑：丁　涵
装帧设计：潘如清
联系电话：0755-82834960

中国林业出版社　·　建筑分社
策　　划：纪　亮
责任编辑：纪　亮　王思源

出版：中国林业出版社
（100009 北京西城区德内大街刘海胡同 7 号）
http://lycb.forestry.gov.cn/
电话：（010）8314 3518
发行：中国林业出版社
印刷：深圳市雅仕达印务有限公司
版次：2016 年 5 月第 1 版
印次：2016 年 5 月第 1 次
开本：235mm×335mm，1/16
印张：20
字数：300 千字
定价：398.00 元（USD 79.00）